Cybersecurity for Reconfigurable Hardware Based Critical Infrastructures

Krishnendu Guha · Jyoti Prakash Singh ·
Amlan Chakrabarti

Cybersecurity for Reconfigurable Hardware Based Critical Infrastructures

Springer

Krishnendu Guha
School of Computer Science
and Information Technology
University College Cork
Cork, Ireland

Jyoti Prakash Singh
Department of Computer Science
and Engineering
National Institute of Technology
Patna, Bihar, India

Amlan Chakrabarti
A. K. Choudhury School of Information
Technology
University of Calcutta
Kolkata, West Bengal, India

ISBN 978-3-031-67590-4 ISBN 978-3-031-67591-1 (eBook)
https://doi.org/10.1007/978-3-031-67591-1

This Springer imprint is published by the registered company Springer Nature Switzerland AG
The registered company address is: Gewerbestrasse 11, 6330 Cham, Switzerland

Contents

About the Authors

Dr. Krishnendu Guha Guha is presently an Assistant Professor/ Lecturer in the School of Computer Science and Information Technology, University College Cork (UCC), Ireland. Prior to this, he was a Post Doctoral Research Associate in University of Florida, USA. He also held the position of an Assistant Professor (Temporary Faculty) in National Institute of Technology, Jamshedpur, India. He even served as a Visiting Scientist in Indian Statistical Institute, Kolkata, India. He completed his Ph.D. from University of Calcutta, India as an INSPIRE Fellow under the Department of Science and Technology, Government of India. He was also associated with Intel India as a Research Fellow. He completed his MTech from University of Calcutta, where he secured the First Class First Rank and was the University Gold Medalist. He has published several research articles in top level journals, conferences and magazines and a book, published by Springer. He works as a reviewer for several IEEE and Springer journals and conferences. Presently, he is a member of IEEE, ACM and IAENG. Presently in Ireland, he heads the QuEST (Quantum and Embedded Security and Technologies) Research Group and is a part of CONNECT and CRT ADVANCE centres of Science Foundation Ireland. His research arena essentially involves Embedded Systems and Hardware Security, Quantum Computing, Real Time Systems Security with a focus on cloud and IoT platforms. He also has a strong interest in Artificial Intelligence and Nature Inspired Approaches.

Dr. Jyoti Prakash Singh is an assistant professor and Head of the Department of Computer Science and Engineering in National Institute of Technology Patna, India. He completed his PhD from University of Calcutta. He has co-authored seven textbooks and one edited book with reputed publishers such as McGraw Hill, Elsevier and Springer. He has published more than 35 international journal publications in leading publishers and more than 50 international conference proceedings. He was involved as an investigator in the MietY sponsored project to develop algorithms for spam calls/fake calls in telephonic conversations. His research interests focus on social media mining, deep learning, information security and speech processing. He is a senior member of IEEE and ACM, Life member of Computer Society of India (CSI) and Indian Society of Technical Society (ISTE) and Institution of Engineers (IE). He is an Associate Editor of International Journal of Electronic Government Research. He was awarded S4DS Data Scientist (Academia) Award by society for Data Science in 2020.

Dr. Amlan Chakrabarti is presently Professor and Director of A. K. Choudhury School of Information Technology (AKCSIT), University of Calcutta. Prior to this, he completed his post-doctoral research at Princeton University after completing his Ph.D. from the University of Calcutta in association with ISI, Kolkata. He has been associated with research projects funded by government agencies and industries related to Reconfigurable Architecture, VLSI Design and Security for Cyber-physical Systems, Internet of Things, Machine Learning, Computer Vision and Quantum Computing. He is the Series Editor of Springer Transactions on Computer Systems and Networks and Associate Editor of Elsevier Journal of Computers and Electrical Engineering. His present research interests include Reconfigurable Computing, Embedded Systems Design, VLSI Design, Quantum Computing and Computer Vision. He has published 200+ research papers in international journals and conferences and has graduated 22 Ph.D. students to date. He is a Distinguished Visitor of IEEE Computer Society and a Distinguished Speaker of ACM (2017–2020).

List of Figures

Chapter 1
Introduction

Objectives of The Chapter

- A brief introduction to embedded processing with reconfigurable hardware.
- Motivation for the current work.
- Summary of chapters for the current book.

1.1 Introduction

Almost every electrical equipment has an embedded processor Konrad and Cheng (2002). While it's easy to spot a processor in smartphones and televisions, this isn't always the case. Appliances in the kitchen, wall-mounted thermostats, and even bank cards are all examples of such gadgets. Actually, embedded applications accounted for nearly all microprocessor manufacture in 1999. Desktops, laptops, and notebooks receive the remaining fraction, which is less than 0.5 percent.

Embedded processors are often domain-specific and designed for certain tasks. Embedded microprocessors are most often used in control applications, digital signal processing, or a combination of the two. Examples of control applications include operating a microwave or a car's cruise control system. Some common examples of digital signal processing are mobile phone video processing and MP3 player audio processing.

The ability of a microprocessor to conserve power is fundamental in numerous contexts. This is often the case when these processors are integrated into energy harvesting or battery-operated devices, or in high-performance computing clusters, where power consumption accounts for a significant portion of data center operational expenses.

K. Guha et al., *Cybersecurity for Reconfigurable Hardware Based Critical Infrastructures*,
https://doi.org/10.1007/978-3-031-67591-1_1

Designing application-specific integrated circuits (ASICs) allows for extremely high energy efficiency, provided that the production volume is sufficiently great. It is more usual to employ off-the-shelf processors, nevertheless, due to the high cost of ASIC creation. These CPUs can run a wide variety of programs, even if they are specialized for their own domain of use. The problem is that this usually means less performance and worse energy efficiency. An ideal processing platform would combine the power savings of an ASIC with the adaptability of a GPU so that it can run a wide variety of applications. In order to run certain applications, this reconfigurable CPU can change its configuration on the fly. Because of this, a variety of applications may run with scalable performance and minimal energy consumption.

The success of Moore's law over the past few decades have redefined the notion of computing Schaller (1997). A variety of new computation platforms have evolved over the ages that has empowered development of a variety of embedded systems associated with diverse functionalities, at very low design costs. Moreover, the exponential growth of embedded devices has been only possible due to the availability of high bandwidth, low latency and cheap network connectivity of present Industry 4.0, especially with the advent of IPv6. This has led to the emergence of a variety of cyber physical systems associated with new application domains.

But as embedded platforms become more diverse and complex, and as we reach the maximum scaling factor of clock frequency on these platforms, we have no choice but to abandon the old methods of designing embedded systems with a single core in favor of more efficient architectures that make better use of hardware while reducing power consumption and thermal emissions. One possible solution is reconfigurable computing.

General purpose computers that run tasks as software programs exhibits high flexibility, but performance is poor. While application specific integrated circuits (ASICs) that can implement tasks as dedicated hardware modules exhibits high performance. However, they are rigid and not flexible and incurs high design cost. Thus, the solution is to have a tradeoff, where performance and flexibility can be combined. Software tasks can be run on a processor, along with hardware tasks on field programmable gate array (FPGA). Moreover, FPGAs possess the ability of dynamic partial reconfiguration at runtime, via which it empowers the host system to execute different hardware tasks at different time instances on the same platform. An example of such a reconfigurable architecture is Xilinx Virtex-4 (XC4VFX60) that comprises of a FPGA fabric and a 32 bit RISC PowerPC processor integrated on the same die.

Programs that execute tasks on general purpose processors are very flexible, but their performance is often subpar. Contrarily, tasks executed as dedicated hardware modules (e.g., ASICs) display high performance but are less flexible and more expensive. A compromise between performance and flexibility, reconfigurable systems enable us to run software tasks on a processor and hardware tasks on an FPGA device, making them a viable option. Field-programmable gate arrays (FPGAs) enhance systems by enabling the simultaneous multiplexing of multiple hardware-based application instances on a single platform.

1.2 Motivation

More and more, sensitive applications are utilizing reconfigurable hardware, including national infrastructures in power grid networks, transportation, aircraft, trains, cars, and military equipment. Security is paramount, and there are major repercussions for failing in any scenario.

Any number of physical or logical methods exist for physically or logically exploiting a computer system, including software and hardware attacks. As reconfigurable hardware grows more powerful and less expensive, attacks can steal sensitive information, alter the system to execute malicious unintended actions, cause a denial of service, or even bring the system to its knees.

It grows more alluring to designers while simultaneously making it easier for attackers to exploit security flaws. Regrettably, reconfigurable hardware's security has had little attention up until now. The need for the use of all various reconfigurable computing platforms in critical infrastructure is utmost essential. Very limited security solutions are existing in literature, and even these are present distributively and are not compiled in a single source. This motivates us to write this book. This book is intended as a first hand course for someone who is interested in developing more secure and trusted reconfigurable hardware based critical insfrastructure.

1.3 Book Structure

The book is structured as follows:

The book commences with an introductory section on reconfigurable computing and thereafter delves into its applications in critical infrastructures. Many reconfigurable computing systems are being deployed to effectively monitor these crucial infrastructures. Security is a crucial concern in computing systems that support critical infrastructures. The book is organized in the following manner:

The present chapter commences with an introduction to embedded computing, especially reconfigurable computing, and emphasises the underlying reasons for authoring this book.

Chapter 2 provides a comprehensive description of the reconfigurable hardware. This chapter provides an introduction to the Field Programmable Gate Array (FPGA), including its architectures, current developments, and diverse applications. This part also covers a compilation of infrastructures that make use of FPGA. The chapter concludes by examining FPGA-based real-time systems designed for both critical and non-critical applications.

Chapter 3 commences with an examination of security administration and planning. This chapter provides an analysis of different security policies, elucidating security vulnerabilities, security measures, and the process of formulating and implementing security policies. This chapter also examines business continuity plans, outlines the methods involved in creating them, and explores their influence on recovery points. This chapter also covers the topics of disaster recovery and emergency response plans. The chapter concludes with the implementation of restoration strategies and the execution of many drills and exercises to reinforce security policies.

Chapter 4 presents the concepts of authentication, authorization, and auditing in the context of reconfigurable hardware-based infrastructure. The text also covers topics such as audit trails, audit mechanisms, clipping levels, compliance audit, authentication procedures, and various configuration management audit authentication techniques. The concluding section of this chapter entails a comprehensive analysis of change management.

Chapter 5 presents the concepts of monitoring and analysis. The text discusses the concept of intrusion detection systems and explores different forms, such as network-based and host-based systems. The technique for creating an IDS alert system is also covered. This chapter discusses different detection and analysis strategies that rely on signature, anomaly, and hybrid modes. This chapter also covers the topics of white listing and black listing in the detection system. Additionally, it concludes by examining the outcomes of specific events, evaluating security metrics, and generating visual representations to effectively communicate these findings to users and stakeholders.

Chapter 6 focuses on the strategies and actions taken to address and mitigate risks, as well as the process of recovering from them. This chapter commences by addressing the concept of risk and delves into the various potential sources of threats, including antagonistic, accidental, structural, and environmental sources. This chapter discusses the many types of threat events and the process of conducting reconnaissance and gathering information in order to eliminate them. This chapter also discusses the susceptibility and management of risk and remaining risk, as well as the identification and evaluation of risk awareness, and the process of reporting and recording. This chapter also covers the process of doing a risk assessment, specifically in relation to quantitative and qualitative evaluations and analysis.

Chapter 7 focuses on the concepts of control and countermeasures. The process begins by doing a cost-benefit analysis and establishing control objectives, which include preventative, detective, and corrective measures. This chapter also discusses additional control measures such as technical security, operational security, and physical security controls. The chapter concludes by discussing various types of backups, including complete backups, incremental backups, and differential backups.

Chapter 8 focuses on the utilization of artificial intelligence and machine learning in the field of cyber security. The chapter discusses the terrain of machine learning within the realm of cyber security. The text also discusses the different elements of machine learning-based systems, various categories of models, and optimizations. This chapter also covers prominent classification techniques, including k-nearest neighbors, logistic regression, decision trees, support vector machines, artificial neural networks, and others. Several unsupervised algorithms, include K-means, hierarchical, and DBSCAN methods. The chapter also explores the integration of artificial intelligence into hardware platforms, such as Field-Programmable Gate Arrays (FPGA) or Application-Specific Integrated Circuits (ASIC).

Chapter 9: This chapter focuses on digital forensics, which explores the realm of cybercrime. It delves into the concept of computers being both targets and tools for cybercriminal activities. This chapter enumerates the many stages of digital forensic inquiry, including acquisition, identification, evaluation, and presentation. This chapter concludes with a discussion on several forms of electronic evidence and their respective categories.

Chapter 10: The upcoming chapter focuses on providing a tutorial and practical experience in hardware system design, attacks, and security. It covers topics such as HDL programming, specifically building a basic VHDL program and Verilog program. This chapter discusses the fundamental gates, components of a system, and finite state machines. The chapter provides comprehensive information on threats that target integrity, confidentiality, and availability in hardware platforms, as well as the corresponding solutions.

References

Konrad S, Cheng BH (2002) Requirements patterns for embedded systems. In: Proceedings IEEE joint international conference on requirements engineering. IEEE, pp 127–136

Schaller RR (1997) Moore's law: past, present and future. IEEE Spectr 34(6):52–59

Chapter 2
Reconfigurable Hardware

Objectives of The Chapter

- Introducing Reconfigurable Systems.
- Overview of FPGAs.
- Infrastructures with FPGAs.
- Real Time Systems and Scheduling.

2.1 What is Reconfigurable Hardware

In the present era, reconfigurable hardware is substantially used in embedded devices Garcia et al. (2006). Reconfigurable hardware is essentially an integrated circuit that comprises several logic gates, whose functionality can be customized at runtime. Even the connections among the logic gates is also configurable. This facilitates execution of several tasks or applications directly on hardware.

The memory cell can be considered as the main element in the fabric of reconfigurable hardware. Such memory cells are essentially in the form of look up tables (LUTs), which is used to implement the universal gates and control the configuration of the switches of the interconnection network. Different tasks are executed directly in hardware via permutation and combination of the logic gates and the interconnection network. The functionality of each gate and the switch state is defined in a program, which is called the configuration file. The basic elements of a reconfigurable hardware are depicted in Fig. 2.1.

In present times, the most common type of reconfigurable hardware device is field programmable gate array (FPGA).

K. Guha et al., *Cybersecurity for Reconfigurable Hardware Based Critical Infrastructures*,
https://doi.org/10.1007/978-3-031-67591-1_2

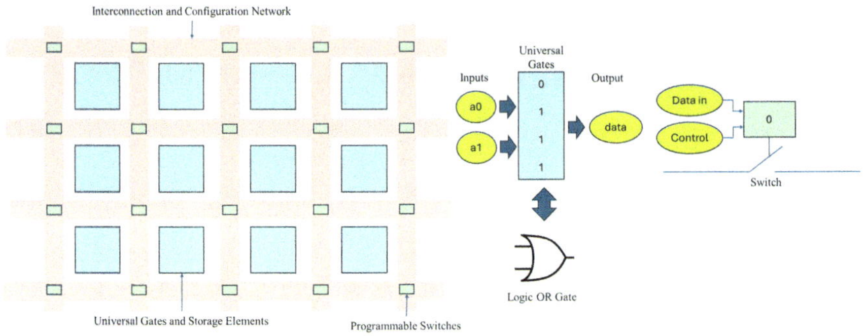

Fig. 2.1 Basic elements of a reconfigurable hardware

2.2 Background to FPGAs

The field-programmable gate arrays (FPGAs) have their origins in PLDs and programmable read-only memories (PROMs) Venkateswaran and Mazumder (1994). But PROMs and PLDs can't be flashed during runtime; that task is reserved for factory or field programming. In addition, once programmed, they are immutable due to the fuse technology. "Burning a PROM" is the result.

In contrast, field-programmable gate arrays (FPGAs) keep their configuration data in a re-programmable device, such as flash memory or static random-access memory (SRAM). So, field-programmable gate arrays (FPGAs) are like integrated circuits—they have user-programmable interconnects and internal hardware blocks. This makes it easier for an FPGA to adapt to new designs, handle new applications, or undergo spatial and temporal reconfiguration. Common companies that make FPGAs include Xilinx, Intel, Microsemi, and others.

So, unlike a microcontroller, we have to write the code and compile it to a hex file before we can load it into the microcontroller. Because it is stored in the flash memory, the software can be updated as needed. But field-programmable gate arrays (FPGAs) require circuit design directly and do not have a processor to run software. Various configurations are possible for the FPGA, ranging from basic gates like AND, OR, and NOT to complex systems like a multicore processor.

The FPGA design process makes use of the hardware description language (HDL). Both Verilog and Very High Speed Integrated Circuit Hardware Description Language (VHDL) are HDL languages that are accessible. Companies use different BITGEN software to convert the HDL file to a bit file. The bit file that sets up the FPGA fabric is this one. When power is turned off, the FPGA file that is stored in the FPGA's RAM is erased. This means that turning on the FPGA requires loading the configuration file every time.

2.3 FPGA Architecture

FPGAs are essentially programmable pre-fabricated silicon chips that are used to implement digital designs Grout (2011). An FPGA architecture essentially comprises of three main components (which are diagrammatically illustrated in Fig. 2.2):

(i) Configurable Logic Blocks (CLBs): used to implement the logic functions.
(ii) Programmable Interconnects or Routing Structures: used for routing signals among the CLBs and implement functions.
(iii) Input-Output (I/O) blocks: provides an interface between the FPGA and other external devices, i.e. used to make off-chip connections.

Let's discuss these in detail.

2.3.1 *Configurable Logic Blocks*

Also termed as programmable logic blocks. This facilitates basic computation and storage of elements in a digital system. Essentially, a CLB comprises of a programmable combinational logic, a flip flop and some fast carry logic for reducing the area, timing delay and also cost. However, modern FPGAs contains much more elements and are a heterogeneous mixture of several blocks like dedicated memory blocks, multiplexers, etc. The configuration memory is used for controlling the functionalities of each element.

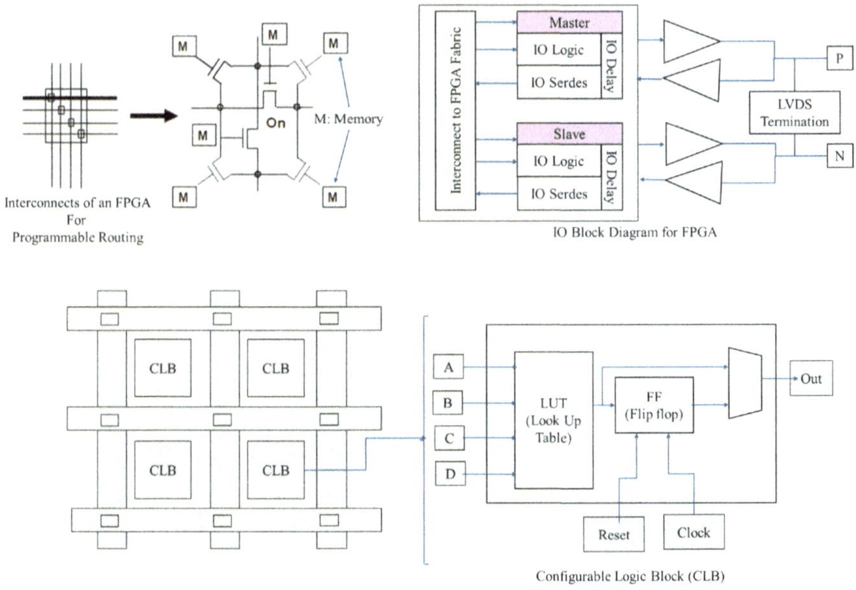

Fig. 2.2 FPGA architecture

2.3.2 *Programmable Routing Elements*

These establishes connections among the CLBs and also with the I/O blocks. It essentially comprises multiplexers, pass transistors and tri-state buffers. The multiplexers and pass transistors are essentially used in a logic cluster to facilitate connection among the logic elements.

2.3.3 *I/O Blocks*

An interface is provided by the programmable I/O pads between the external components and the intrinsic FPGA components, i.e. the logic blocks and routing architecture. With the surrounding logic circuit, the I/O pad forms an I/O cell or I/O block. Design of the I/O blocks is quite complex as they have to handle great differences in the supply voltage and reference voltage. Moreover, selecting the standards is important for the I/O design, as supporting a large number of standards increases the silicon chip area required for the I/O cells.

Modern FPGAs are used in a variety of applications and hence several special functional blocks are added along with the basic ones. The special functional blocks include ALUs, block RAM (BRAM) cells, multiplexers, digital signal processing elements, microprocessors, etc.

2.4 Design Flow

Designing a hardware circuit in an FPGA involves the following steps:

(i) design entry;
(ii) design synthesis;
(iii) design implementation;
(iv) device programming;
(v) design verification.

A diagrammatic representation of the design flow is depicted in Fig. 2.3.

2.4.1 *Design Entry*

Introducing a hardware circuit in an FPGA can be done in a variety of ways. The two most common techniques are (i) HDL based design entry and (ii) schematic based design entry.

In the former case, a HDL program (either in VHDL or Verilog) is written that describes a hardware circuit. In the latter case, a graphical user interface (GUI) is

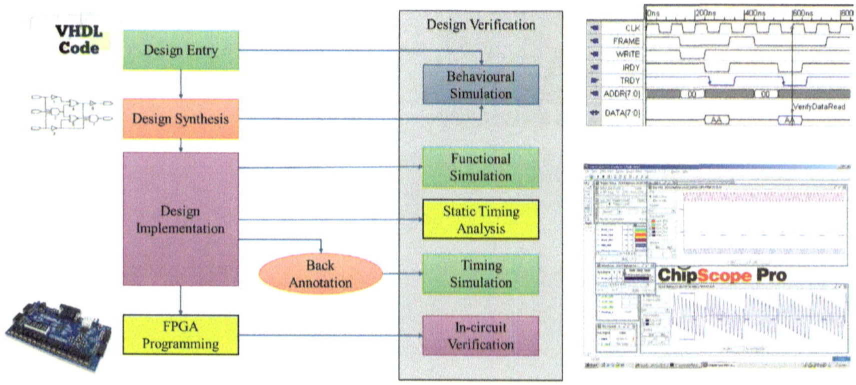

Fig. 2.3 FPGA design flow

used to design the circuit, whose basic elements or components are selected from a design library.

While the schematic based entry supports greater visibility and control over the hardware components, the HDL based technique facilitates the designer to transfer algorithmic types of concept to hardware circuits. Hybrid mode of design technique is also possible, where both the HDL mode and schematic mode can be combined for designing.

2.4.2 Design Synthesis

In this step, the entered design is converted into a device netlist, which is essentially a complete circuit with logical elements. This design synthesis process essentially checks the code syntax if the design is entered in the HDL format and also analyses the hierarchy of the entered design architecture. In this step, the design is also optimized if necessary. Finally, the netlist is saved in Native Generic Circuit format or simply, as a NGC file.

2.4.3 Design Implementation

The next step is design implementation, which essentially translates the entered design into basic components and places and fits them into the CLBs of the FPGA. This step also defines how communications or routing of signals would take place between the CLBs of the FPGA to ensure working of the entered design. Essentially, this phase comprises three steps: (a) Translate (b) Map (c) Place and Route. Details of these three steps is shown diagrammatically in Fig. 2.4.

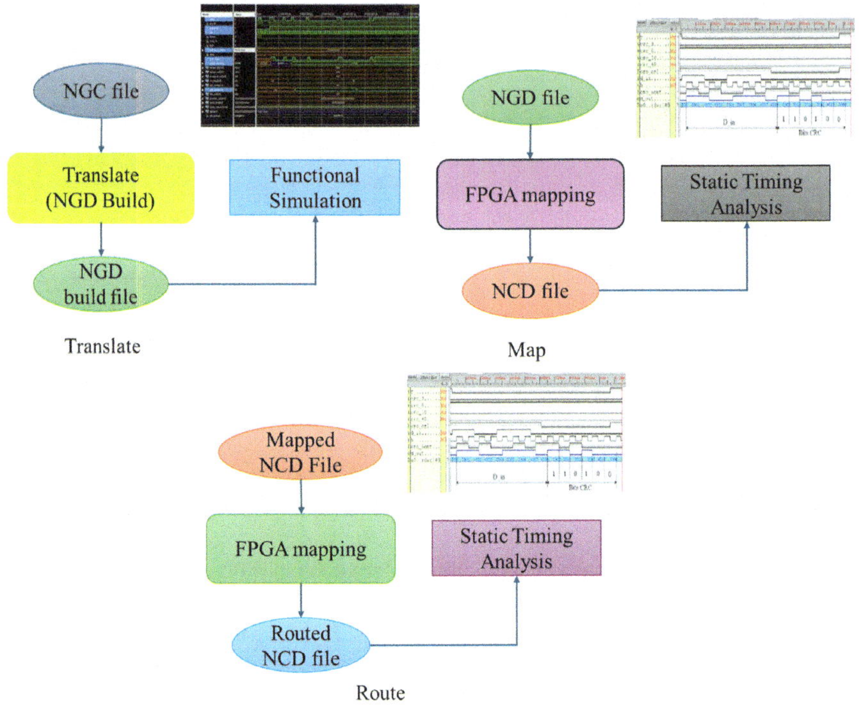

Fig. 2.4 Translate, map and route

2.4.3.1 Translate

In the translate phase of design implementation, all the input netlist files are combined to the logic design file that is saved as a Native Generic Database (NGD) file. Functional simulation can also be performed with the generated NGD file. In this phase, all the ports of the design are assigned to the actual FPGA elements like the pins and switches. This configuration is essentially stored in a user constraints file (UCF).

2.4.3.2 Map

In the mapping phase of design implementation, the design circuit that is to be mapped to the FPGA fabric, is divided into several sub-blocks that can be placed in the FPGA logic blocks or CLBs. Hence, the NGD file is transformed into a NCD file, which essentially contains which portion of the desired design is to be mapped to which portion of the FPGA. Static timing analysis for the design sub-blocks mapped into appropriate logic elements of the FPGA can be done at this place.

2.4.3.3 Routing

In the routing phase, communication among the design sub-blocks placed into the appropriate CLBs of the FPGA are defined. Design rule checks are also performed. Thus, from the mapped NCD file, a routed NCD file is generated. Static timing analysis of the routed design of the FPGA can be performed at this stage.

2.4.4 Device Programming

After design implementation, the routed design must be loaded or physically placed on the FPGA. For this to be enacted, the design needs to be converted in a format which the FPGA can understand and support its placement. Hence, the routed NCD file needs to be converted into a BIT file, which is performed by the BITGEN program. Th BIT file can configure the FPGA fabric.

2.4.5 Design Verification

This essentially comprises of functional verification and timing verification that can be done at various stages of the design flow. The following types of verification are associated.

2.4.5.1 Behavioral Simulation

This is also known as RTL simulation and is the first of all the verification stages that is allowed in the hierarchy of the design flow. This can be performed before the synthesis stage and is usually done to verify the RTL code. Essentially, correctness of the signals and variables are observed and the procedures and functions of the RTL design are traced and breakpoints are set at this stage.

2.4.5.2 Functional Simulation

This is performed after translation of the RTL design. Validity regarding the logical operation of the circuit is provided.

2.4.5.3 Static Timing Simulation

After mapping of the design, static timing simulation can be performed. The post map timing report that is generated at this stage provides the signal path delays. If static timing analysis is performed after the place and route phase, the timing report that is generated even considers the timing delay information. A complete timing summary of the design is generated.

2.5 FPGA Trends

The gap between discrete logic devices and the entry level FPGAs are filled by less complex devices like simple programmable logic devices (SPLDs) and complex programmable logic devices (CPLDs). While the SPLDs comprises of some flip flops and multiplexers, the CPLDs contains some logic blocks and an interconnection matrix with I/O ports. The basic structure of a SPLD is depicted in Fig. 2.5, while that of a CPLD is shown in Fig. 2.6.

The entry level FPGAs essentially emphasize on less power consumption and low logic density and complexity per chip. However, present day high function FPGAs include additional functional blocks that are dedicated to specific functionalities. This includes dynamic clock management (DCM) modules, phase locked loops (PLLs) and high speed serializers and de-serializers. Even high-speed communication components like Ethernet MACs, PCI express controllers and high-speed transceivers are integrated into the FPGA device. Integration of such blocks is either done with the CLBs of the FPGA, which are then termed as soft IPs or integrated as separate

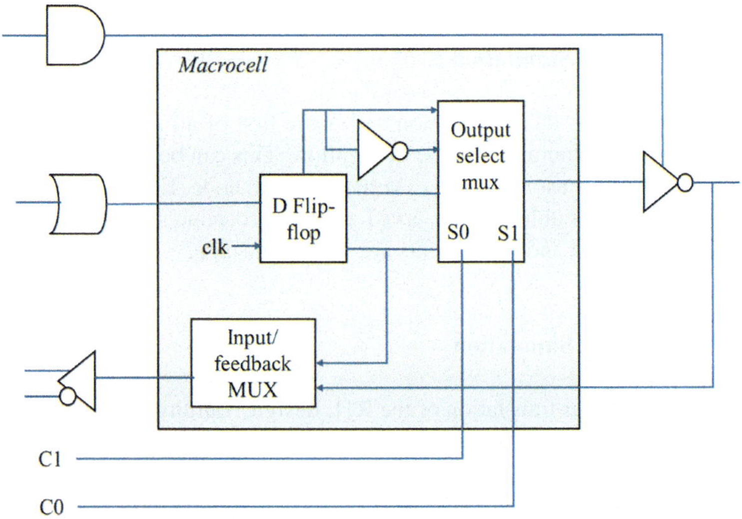

Fig. 2.5 Structure of SPLD

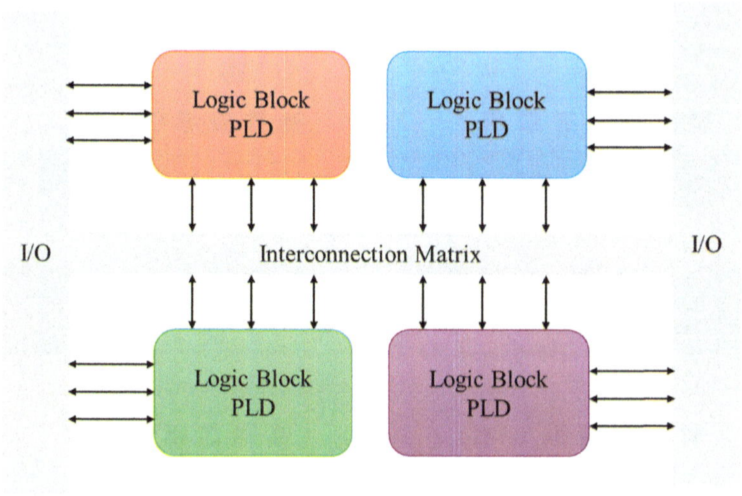

Fig. 2.6 Structure of CPLD

circuits, commonly known as hard IPs. At the expense of reconfigurability, the hard IPs gain performance.

Present-day high-end FPGA boards include complex system on chip (SoC) components that integrates the FPGA architecture, hard IP and a microprocessor CPU core into a single device. When compared to separate devices, the SoC FPGA offers better integration, less power, small board size and high bandwidth communication between the core and other blocks.

Lets consider Intel's Cyclone V, a typical example of SoC FPGA. This can be used for high volume applications as in industrial motor control drives, video processing applications, in network applications like protocol bridging and even in handheld devices. The basic architecture is shown in Fig. 2.7, which has two distinct divisions. One is the FPGA portion and the other is the hard processor system (HPS) that is based around a single or dual core 32 bit Arm Cortex A9 MPCORE, which runs at 925 MHz. Each part has its own peripherals that even includes hard IPs from third party vendors. Similarly, the Stratix 10 SX that comprises an FPGA with a quad core 64 bit Arm Cortex A53 running at 1.5 GHz, is targeted for high performance applications like communication systems, hardware acceleration for data centres, high performance computing (HPC), radar processing and even for ASIC prototyping.

Moreover, with the facility of high level synthesis (HLS), system developers can precisely configure the FPGA with high level languages like C, C++, System C and OpenCL.

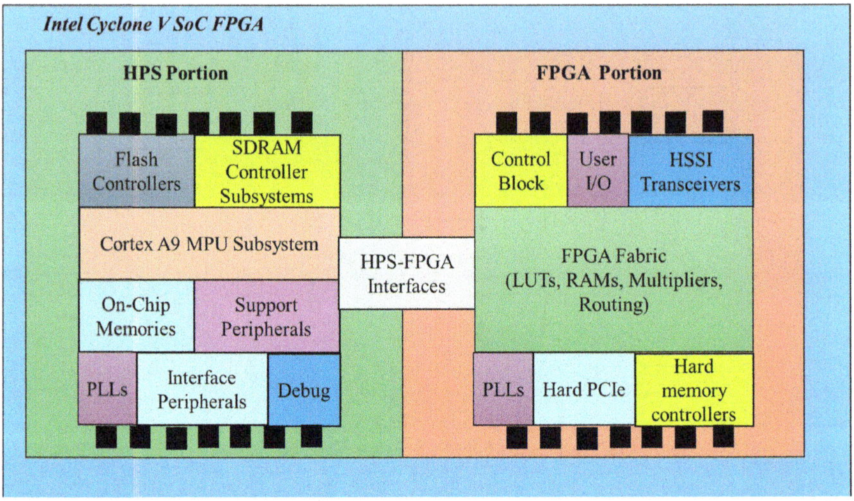

Fig. 2.7 Structure of FPGA SoC

2.6 FPGA Applications

Embedded FPGA architectures has great opportunities in various industrial sectors due to the growing demand of high bandwidth at low cost and low power. According to a report by Market Research Future, it was seen that the worth of FPGA market in 2018 was worth to be USD 9 billion and it was estimated that it would develop at a Compound Annual Growth Rate (CAGR) of 9.7% from 2020 to 2027.

FPGAs are used in various arenas and its usage is expected to grow in recent future as its usage is being adopted in several important sectors like aerospace and military applications. Other than these, FPGAs are also used in consumer electronics, manufacturing, automobiles and wireless applications. In such applications, FPGAs are used for waveform generation, image processing, communication, security, network processing and deep packet inspection. Other than this, FPGAs are well used in academia and research sectors for experimentation and analysis. Figure 2.8 [*Source: grandviewresearch.com*], depicts the global FPGA market share in various applications.

Other than standalone environments, FPGAs are widely used in cloud and edge platforms. They are mostly used as hardware accelerators in cloud and edge platforms. Such FPGA based accelerators serve as strong contenders to conventional GPU based accelerators for high performance computing in cloud and edge platforms.

In general, FPGAs are mostly used in cloud platforms for hardware acceleration. Cloud service providers tend to speed up time consuming tasks like network encryption, deep learning, memory caching, webpage scoring, high frequency trading, video conversation. Other common use of FPGAs involve evaluation of neuromorphic computing applications like deep neural network based applications, acceleration of software defined networking. Even Bing search results are ranked with the use of FPGAs.

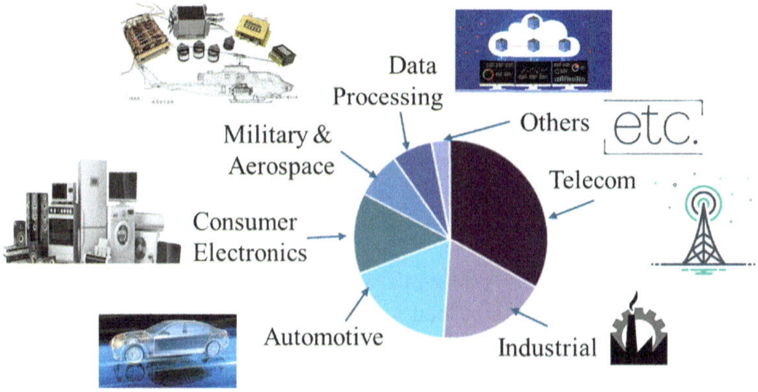

Fig. 2.8 Market share of FPGA as per application sector

Moreover, cloud service providers is providing hardware as a service (HaaS) to customers, in addition to the already existing software as a service (SaaS), platform as a service (PaaS) and Infrastructure as a service (IaaS). In HaaS, users or customers can access the services of the FPGA resources placed in the cloud, in return of service fees, without physically owning the FPGAs. The users can either send custom generated programs that are taken up by the cloud service providers as image files (a type of encrypted bitstream) that is used to configure the FPGAs placed in the cloud or the users can use pre-generated bitstreams placed in the cloud that are procured from third party vendors. Charges for using the FPGAs or the service fees depend on the type of FPGA chosen and the type of work that is intended to execute in the FPGA. Practical scenarios include Amazon's EC2 F1 services and Microsoft's Project Catapault.

With introduction of HaaS and existing use of FPGAs for hardware acceleration in cloud and edge platforms, the market share for FPGAs is increasing rapidly.

2.7 Critical and Non-critical Applications with FPGAs

As studied in the previous section, FPGAs are used in a variety of applications. Such applications may be either critical, non-critical or mixed critical in nature. In particular, a system infrastructure has several components that executes different types of functions or tasks, that can be defined as critical or non-critical, based on its nature and importance Rodríguez-Andina et al. (2015).

2.7.1 Critical

Critical components of a system infrastructure are associated with tasks that need accuracy of results and must be completed within strict time instant. Failure may cause fatal incidents or even catastrophic results. In general, hard real time tasks that are discussed in consequent sections are critical in nature. For instance, braking system of an automobile can be considered as critical.

2.7.2 *Non-critical*

Non-critical components of a system infrastructure neither need result accuracy, nor are time dependent. These do not lead to catastrophic results and can be redone. In general, soft real time tasks that are discussed in consequent sections are non-critical in nature. For instance, music system of an automobile can be considered as non-critical, as failure of such an infrastructure will only affect mood of the user, but will not cause any fatal incidents.

2.7.3 *Mixed Critical*

When a processing resource is used for executing different types of tasks and services multiple users, then the tasks have to be classified as critical and non-critical or ordered based on their level of criticality.

For instance, processing resources in cloud and edge platforms caters to a number of users. A designer must understand the nature of tasks and categorise them as critical and non-critical and allot resources accordingly. High priority is generally given to critical tasks, while low priority is given to non-critical tasks, while scheduling them for execution. Similarly, the cloud service providers charge a high cost for execution of high critical tasks and even provides better security during their execution.

2.8 Infrastructures with FPGAs

2.8.1 *Standalone Environments*

These are normally used for academic and research purposes. Training and experimentation is generally performed, before actual deployment. A descriptive diagram is shown in Fig. 2.9.

Bitstreams or reconfigurable intellectual properties (RIPs) that configure the FPGA fabric to execute the tasks are stored in a repository. These RIPs may be procured from multiple third party vendors. Based on the user preference, a suitable RIP is chosen from the repository to configure the related FPGA and execute the task. This can have two modes of operation: (i) Dedicated (ii) Shared. These are depicted diagrammatically in Fig. 2.10.

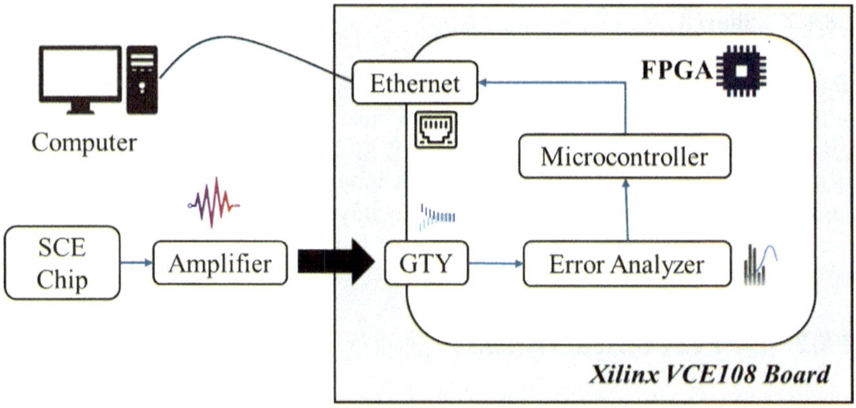

Fig. 2.9 FPGA use in standalone environment

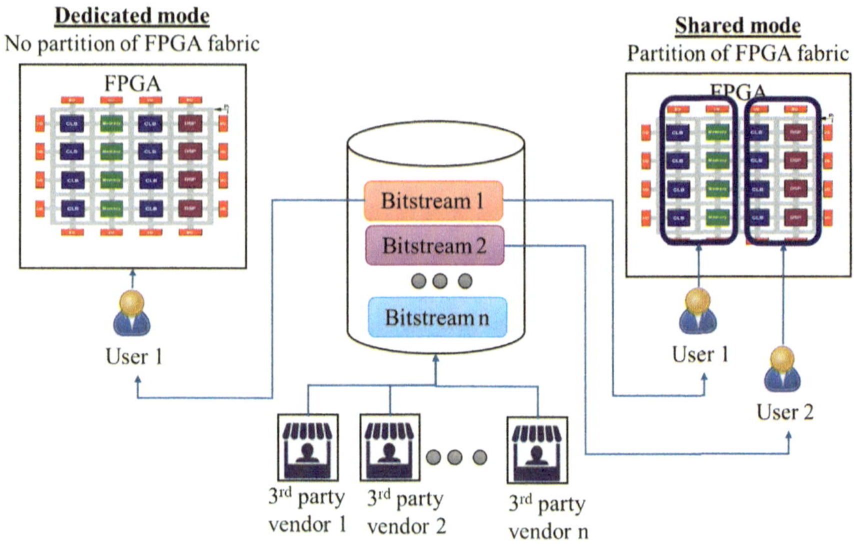

Fig. 2.10 FPGA modes of operation

2.8.1.1 Dedicated

In dedicated mode of operation, the entire FPGA is used to execute a dedicated user task. For this, the entire FPGA fabric is configured by a single bitstream or reconfigurable intellectual property (RIP), associated with the execution of the task. Multiple RIPs may configure the FPGA at different instances of time. Thus, temporal scheduling can be generated.

2.8.1.2 Shared

In shared mode of operation, the property of dynamic partial reconfiguration (DPR) of the FPGA is exploited. In this, multiple virtual portions (VPs) are created. Each VP can be configured with a dedicated RIP. Multiple user tasks can be executed in multiple VPs of the FPGA. At different time instances, different VPs can be configured with different RIPs. The DPR facility of FPGAs is helpful for spatio-temporal mode of scheduling and execution of multiple user tasks.

2.8.2 e-FPGA Based Systems

Present day system on chip (SoC) platforms has FPGAs integreted in their fabric, as shown in Fig. 2.11. In addition to the FPGA, there comprises an ASIC element and a lightweight processor like microblaze or ARM. Intercommunication of data may take place between the softcore processor, FPGA and ASIC elements. There can be multiple e-FPGAs on an SoC. Along with these, the SoC may comprise of embedded memory blocks, plase locked loops, digital clock management blocks and communication modules.

Most of the embedded devices comprise of such SoC platforms with e-FPGAs. These are normally found as the end-user devices.

Like the previous case, the FPGAs can be utilised either in a dedicated mode or in a shared mode. Temporal type of scheduling is associated in dedicated mode, while the shared mode offers spatio-temporal style of scheduling and execution.

2.8.3 Parallel and Distributed Processing Systems

Present day IoT platforms have multiple end user devices communicating with each other, with the objective of delivering quick inference and ensuring security

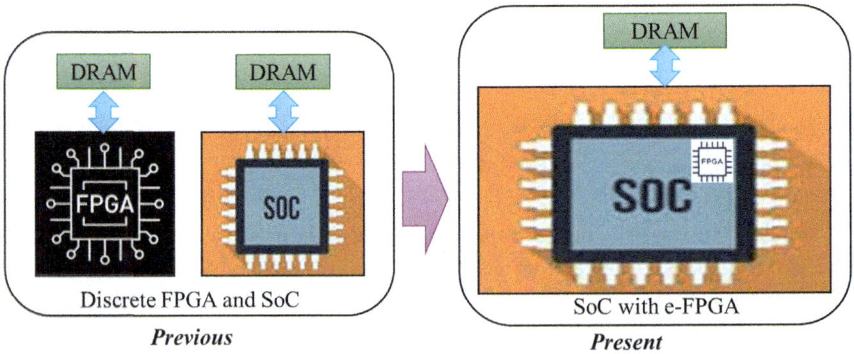

Fig. 2.11 Move to SoC with e-FPGAs

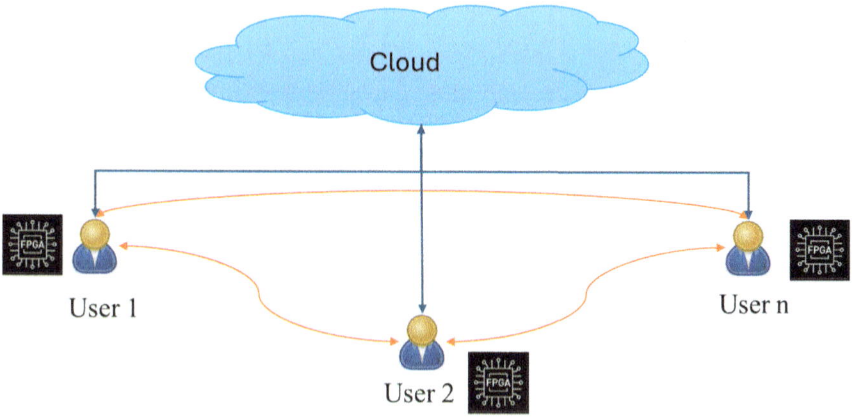

Fig. 2.12 Deployment of FPGAs in parallel and distributed processing systems

without moving back to the edge and cloud platforms for computing. A diagrammatic representation of the architecture is shown in Fig. 2.12.

Such end devices generally comprise of SoCs that have a single or multiple FPGAs. Such devices may transfer a task or share the operation of tasks among themselves. The key thing is that such IoT nodes generally possess an agent that handles the real time controls and decision makings. These agents may be simple software programs directly loaded to the SoC platform or may be additional hardware plug and play entities.

These agents select requests and tasks that can be accommodated and executed on its host platform from all the incoming tasks and requests coming from other platforms and users.

2.8.4 Cloud and Edge Infrastructures

The current era has witnessed cloud service providers (CSPs) delivering Hardware as a Service (HaaS), along with Software as a Service (SaaS), Platform as a Service (PaaS) and Infrastructure as a Service (IaaS). In general, HaaS features firmware updates, with commissioning software updates, servers spaces for several activities like recording via cameras in car, capital operational expenses, trained installers and tech support, etc.

Even FPGAs are deployed in the cloud platforms. A descriptive diagram is presented in Fig. 2.13. Some well known examples are like IBM (PureApplication), Oracle (Exa-Data), Cisco (UCS), Microsoft (Azure), Baidu cloud, and Google (Compute Engine). Normally CSPs use FPGAs as a hardware accelerator for user tasks. Along with these, the CSPs offer users to use their on-cloud FPGAs to execute their custom applications, in return of certain fees or via a pay per use mechanism. Users

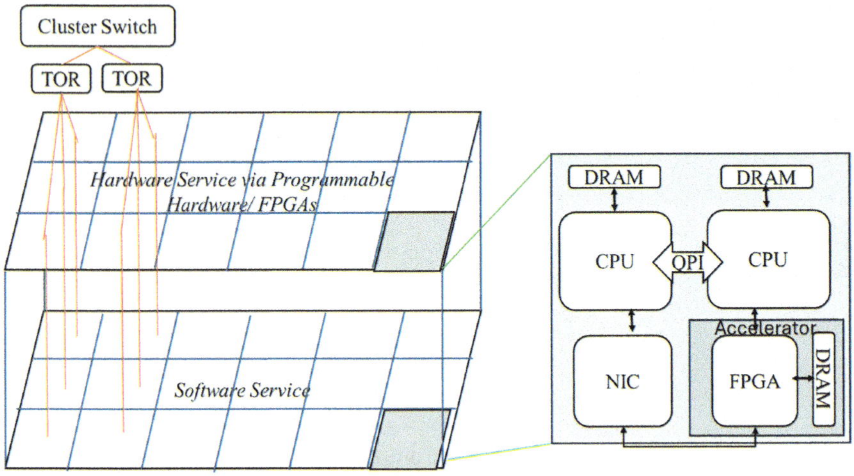

Fig. 2.13 Deployment of FPGAs in cloud and edge platforms

no longer need to own FPGAs, to execute their custom applications on an FPGA. Practical examples include Amaon's EC2 F1 services, Microsoft's Catapault.

Other than cloud platforms, FPGAs are even deployed in the edge platforms. This is instantiated to facilitate quick computation near to user ends.

Such architectures comprise of a layer of FPGAs and a layer of CPUs. The CPUs are interconnected via an interconnection bus. The CPUs are also connected to an FPGA cluster in the FPGA layer via PCI-e interfaces or any standard cable connection. Each FPGA cluster may consist of a single FPGA or multiple FPGAs. When a cluster comprises multiple FPGAs, then one acts as the master (which is directly connected to the CPU), and other acts as slaves. The FPGAs may also be connected to each other (among different clusters) via shared memories. These serve as excellent accelerators and even can share tasks and execute them in parallel, when higher efficiency is required.

2.9 FPGA Based Real Time Systems

2.9.1 Real Time Systems

The physical passage of time determines how a real-time system evolves. The controlled object, the real-time computer system, and the human operator are the three main components of a real-time system. The controlled object or operator must provide input to a real-time computer system within the time intervals specified by the system's environment. A number of fields make use of real-time systems, including aerospace, defense, embedded automative electronics, multimedia networks, and

Fig. 2.14 Types of real time systems

more. The accuracy of a real-time system's actions is dependent on both the timing of the computations and the logical outcomes of those computations. A deadline is the exact moment when a product is ready for use. Figure 2.14 shows a visual depiction of the different kinds of real-time systems.

2.9.1.1 Soft Real Time Systems

If the result has utility even after the deadline has passed, the deadline is classified as soft.

2.9.1.2 Firm Real Time System

A firm real-time system can be defined as one where a few missed deadlines does not lead to total system failure. It lies in the middle of hard and soft real time systems.

2.9.1.3 Hard Real Time Systems

A catastrophe could result if a firm deadline is missed, the deadline is hard. Examples of hard real time systems: Commands and Control systems, Air traffic control systems. Examples of soft real time systems: On line transaction systems, airline reservation systems.

2.9.1.4 Differences Between Soft and Hard Real Time Systems

Failure to meet the millisecond-or less response time requirements of hard real-time systems can have catastrophic consequences. The response time requirements for soft real time systems, on the other hand, are higher but not as strict.

Peak load performance in a hard real-time system needs to be predictable and stay within the set deadlines. It is possible to tolerate impaired operation under infrequent peak loads in a soft real-time system.

No matter what, a hard real-time system has to always be in sync with the environment. But under heavy load, soft real-time systems' response times will degrade.

Safety is of the utmost importance in hard real-time systems. Data files and databases used by hard real-time systems are relatively small. Here, precision in time is frequently of the essence. A longer-term commitment to real-time system integrity is necessary for soft real-time systems, such as online reservation systems, which use larger databases.

When a soft real-time system encounters an error, it will begin the recovery process by reverting the computation to a previously set checkpoint. Rollback and recovery are not very useful in hard real-time systems.

2.9.2 *Nature of Tasks*

There is a real-time requirement that the system respond within a certain amount of time or meet a certain deadline. Systems for flight control, real-time displays, etc.

Periodic tasks and dynamic tasks are the two main categories of work in real-time systems.

A pictorial representation of the nature of tasks is shown in Fig. 2.15.

2.9.2.1 Periodic Tasks

Periodic tasks involve releasing jobs at regular intervals. A periodic task is one that repeats itself after a set time interval. Four tuples represent a periodic task.

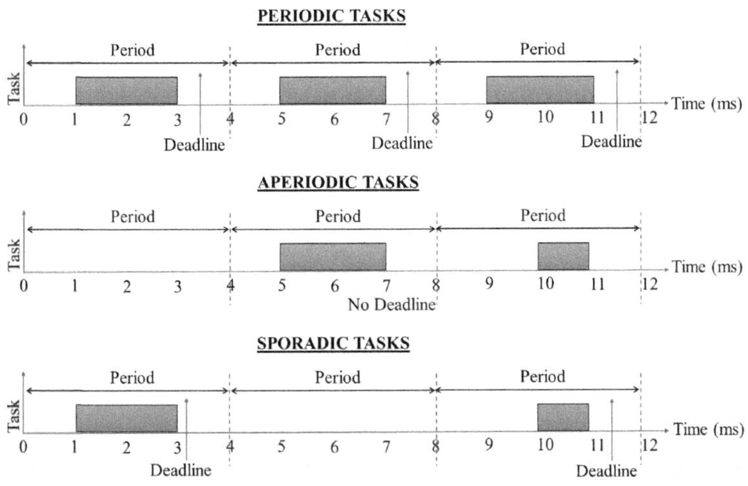

Fig. 2.15 Nature of tasks

$T_i = \langle \phi_i, P_i, e_i, D_i \rangle$ Where ϕ_i represents the task's phase. The phase represents the release time of the task's first job. If the phase is not specified, the release time for the first job is assumed to be zero. P_i is the task's period, or the time interval between two consecutive job releases. e_i i represents the task's execution time. D_i represents the task's relative deadline.

For example: Consider the task: T_i with period = 5, execution time = 3. Because the phase is unknown, we can assume that the first job's release time is 0. So the job of this task is first released at t = 0, where it executes for 3 seconds, and then the next job is released at t = 5, where it executes for 3 seconds, and so on until t = ten. So, jobs are released at t = 5k, where $k = 0, 1, ..., n$.

The hyper period of a set of periodic tasks is the smallest common multiple of all the tasks in the set. For example, two tasks T_1 and T_2 with periods 4 and 5, respectively, will have a hyper period, H = $\text{lcm}(p_1, p_2) = \text{lcm}(4, 5) = 20$. The hyper period is the point at which the pattern of job release times begins to repeat.

2.9.2.2 Non-periodic Tasks/Dynamic Tasks

An event triggers the execution of this sequential program. Both internal and external processes within the system have the potential to produce an event. Criticality and occurrence time knowledge allow for the categorization of dynamically arriving tasks.

Periodic Tasks: Jobs are released in this type of task at random time intervals. Job deadlines for aperiodic tasks are either very flexible or nonexistent at all.

Sporadic tasks: occur at random intervals, they are comparable to aperiodic tasks. Sporadic tasks have strict deadlines, which is the only distinction. Three tuples signify a sporadic task. In each iteration, T_i = 'e_i, g_i, D_i' In which e_i is the task's execution time. g_i is the smallest potential interval between two successive executions of the task. The relative due date of the task is D_i.

2.9.3 *Real Time Task Scheduling on FPGAs*

2.9.3.1 Scheduling

Assigning resources to carry out tasks is known as scheduling in computing. Processors, network connections, and expansion cards are all examples of resources. Data flows, threads, or processes might be the tasks at hand. A process known as scheduler is responsible for carrying out the scheduling activity. The goal of most scheduler designs is to:

- engage every available computer resource (like load balancing),
- make it possible for numerous users to efficiently share system resources,
- reach a desired level of service quality.

A computer system's execution model includes scheduling, which is essential to computation. It is feasible for computers to multitask using a single central processing unit thanks to the idea of scheduling.

2.9.3.2 Goals of a Scheduler

A scheduler may aim at one or more goals, for example:

- getting the most done in the least amount of time possible (maximize throughput),
- cutting down on wait time (the interval between when work is prepared and when it starts),
- minimizing latency or response time.
- giving each process an equal amount of CPU time, or more generally, times that are suitable based on each process's priority and workload, (achieving fairness).

2.9.3.3 Scheduling Policy/Algorithm

An algorithm for allocating resources among parties that asynchronously request them is known as a scheduling discipline, scheduling policy, or scheduling algorithm. Simple scheduling algorithms include FIFO, EDF, SJF, etc., which stands for first in, first out. These are applicable for scheduling systems in real time and are common in many operating system scheduling strategies.

Executing a set of real-time tasks concurrently in a way that guarantees all time-critical tasks meet their deadlines is the hallmark of a hard real-time system. Computational and data resources are essential for the completion of any task. Allocating resources to meet time constraints is the focus of the scheduling problem.

Static and dynamic real-time scheduling are the two main categories.

Make your scheduling decisions at compile time with static scheduling. Using what is known about the parameters of the task sets (such as deadlines, maximum execution times, precedence constraints, and mutual exclusion constraints), an offline schedule is created for run time. For this reason, the overhead during runtime is minimal.

It is during runtime that dynamic scheduling decides which of the available tasks to schedule. The dynamic schedulers can change and adapt to new situations. However, run time processing can cause them to incur substantial overheads. Both static and dynamic scheduling allow for the scheduling of tasks in either a preemptive or non-preemptive fashion.

In preemptive scheduling, a higher-priority task will take precedence over the one that is currently running.

A non-preemptive scheduler will wait for the currently running task to finish before preempting it.

2.9.3.4 Scheduling for Reconfigurable Hardware

For reconfigurable hardware, a spatio-temporal scheduling strategy is followed. This means that multiple users are catered by dividing the resource with respect to space and time. Dynamic partial reconfiguration strategy is followed to achieve this Guha et al. (2021).

For spatial scheduling, the FPGA or reconfigurable hardware fabric is partitioned into multiple parts or virtual partitions (VPs), as described before and each is used to execute a task of a different user. This is performed via the shared mode of operation as discussed in Sect. 2.7.1.2 and shown in Fig. 2.10.

For temporal scheduling, different user tasks are scheduled to execute in the various FPGA VPs at different instants of time. Though the standard modes of scheduling can be used, however, designers prefer to customize their scheduling strategy as per the scenario.

Note: FPGAs has a single ICAP port, hence, only one VP can be reconfigured at a time. For each task execution, there is a reconfiguration time that is associated with reconfiguring the FPGA VP with the related bitstream and then the task execution time.

Questions

1. Explain how configurable logic blocks in an FPGA fabric helps to execute different types of task execution.
2. Describe the differences between design synthesis and design implementation when configuring a bitstream on an FPGA device.
3. State the intrinsic details of a SPLD and CPLD.
4. Consider an automobile system. Justify with reasons, which functions of the automobile system can be considered as critical and which as non-critical.
5. Describe how a single FPGA can be used to cater to multiple users at the same time.
6. What are e-FPGA based systems? How they differ from standalone FPGAs?
7. Explain how FPGAs can be used in parallel and distributed infrastructures.
8. What is HaaS? State how FPGAs can be used for HaaS.
9. Explain with examples the difference between hard real time systems and soft real time systems.
10. Differentiate between periodic and dynamic tasks with examples.

References

Garcia P, Compton K, Schulte M, Blem E, Fu W (2006) An overview of reconfigurable hardware in embedded systems. EURASIP J Embed Syst 2006:1–19

Grout I (2011) Digital systems design with FPGAs and CPLDs. Elsevier

Guha K, Saha S, Chakrabarti A (2021) Self aware security for real time task schedules in reconfigurable hardware platforms. Springer. https://doi.org/10.1007/978-3-030-79701-0
Rodríguez-Andina JJ, Valdes-Pena MD, Moure MJ (2015) Advanced features and industrial applications of FPGAs-a review. IEEE Trans Ind Inform 11(4):853–864
Venkateswaran R, Mazumder P (1994) A survey of DA techniques for PLD and FPGA based systems. Integration 17(3):191–240

Chapter 3
Security Administration and Planning

Objectives of The Chapter

- Understanding Security Policies and their Development.
- Security Administration and Planning.
- Understanding Business Continuity Plan.
- Implementing BIA, understanding its Output and Impact Analysis.
- Understanding Disaster Recovery Issues.
- Emergency Response Planning and Management.

3.1 Security Policies

3.1.1 What is a Security Policy?

A security policy can be described as a written document that provides a high level view about the security goals of an organization or a service provider Harris (2010). Such a document also specifies the requirements needed for an existing process or technology and even provides the requirements that would be needed to introduce a new process or technology. This is depicted in Fig. 3.1.

From employee perspective, such a document provides a guideline for the employee. Though it may not be mandatory, but in general, such a document provides the list of actions that an employee is to perform in different situations. Thus, it guides an employee and directs him/her regarding his/her work and functionality, which will ultimately aide the organization to achieve its security goals.

From user perspective, the user knows what type of security it can get as a service from the company or organization.

K. Guha et al., *Cybersecurity for Reconfigurable Hardware Based Critical Infrastructures*,
https://doi.org/10.1007/978-3-031-67591-1_3

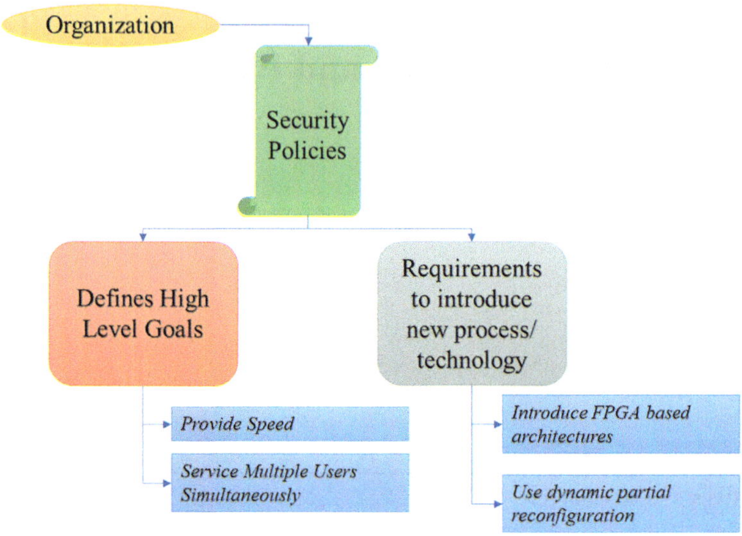

Fig. 3.1 Security policy basics

From the perspective of reconfigurable hardware based infrastructures, it is of particularly important as the reconfigurable hardware fabric is configured with different bitstreams at different instances of time. Hence, it is quite important to understand how the security policy is uphold when the bitstreams are changed and allocated to different users at different instants of time.

Even the security policy for handling critical tasks is different from that of non-critical tasks. Thus, efficient security policies must be developed for a system that is handling mixed critical tasks in different portions of the overall system architecture and supporting different users, with different objectives.

3.1.2 Security Policy for Risk Management

Risks may arise with time. Be it associated with the base hardware of the system infrastructure or the top-level application softwares that are running in the system. Or it may be associated with user policies as a loophole in user policies will provide the liberty to an user for foul play. Even an organisation may be at risk of losing its customers if it does not update its policies with time, depending on Government rules and advisories, or if its competitors provide some offers that may lure its customers Fenz et al. (2014).

For standalone systems, manual supervision may be sufficient for risk management. However, for parallel and distributed systems and systems associated with

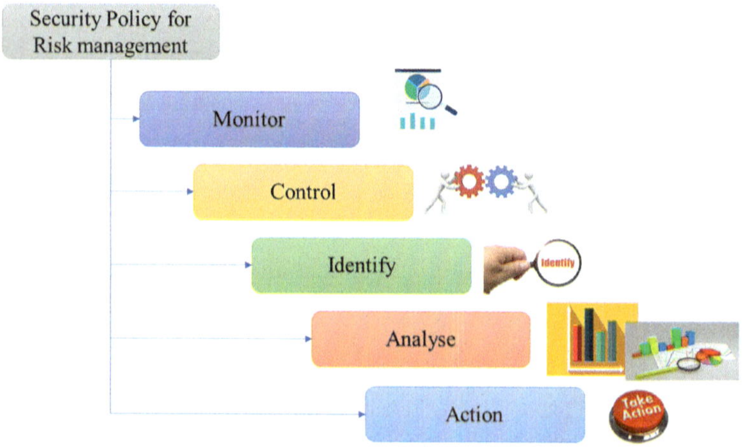

Fig. 3.2 Security policy for risk management

cloud and edge platforms, agent based management is needed, which works based on a security policy.

The main aspects of a security policy (that involves risk management) are depicted in Fig. 3.2 and are also enumerated as follows:

(i) Monitor: This phase involves monitoring or observing the operations of system or activities done by the user.
Hardware agents that monitors the reconfigurable hardware needs to provide more attention to sections that are associated with execution of critical tasks. Normally, these monitor the timing parameters, but other side channel parameters like power, may be also focussed on.

(ii) Control: This phase is associated with controlling the operations of the system and preventing it to move into some unknown state.
The hardware agents on detecting any deviation from the normal behaviour needs to control the operations of the system to prevent it from moving into some anomalous state. For this, it needs to identify the anomaly and categorise it.

(iii) Identify: This phase is associated with identification of risk that may occur with time.
Hardware agents are fed with some learned parameters. These are developed based on supervised and unsupervised learning techniques. These tries to identify the anomaly, like whether the deviation was unintentional like fault or was caused intentionally as due to hardware trojan effect.

(iv) Analyse: This phase involves analysing the possible risk mitigation strategies that are present.
Hardware agents on identifying problem, needs to analyze what will be the possible solutions or possible techniques that can avoid the risk.

(v) Action: This phase is associated with selecting a suitable risk mitigation strategy and applying it to bring back the system into a risk free zone.

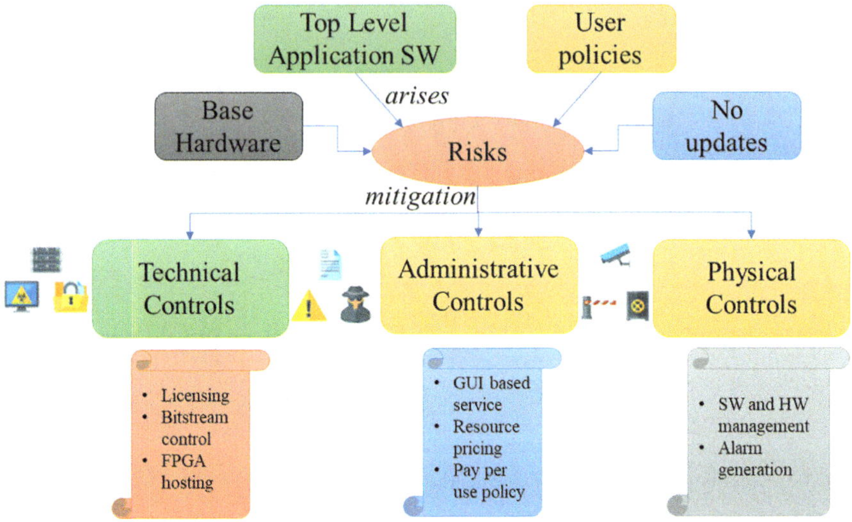

Fig. 3.3 Risk management

Based on the analysis and shortlisting of the possible mitigation scenarios, the hardware agent selects a line of action and applies it to restore the normal scenario. This can be associated with possible rollback, or outsourcing a task to another platform, or even reconfiguring the entire system with another bitstream.

Risk management technique is depicted in Fig. 3.3.

3.1.3 Security Controls

In case of encountering a risk, different control strategies must be applied to mitigate the risk. These are also enlisted in the security policy.

In general, a security policy may comprise of three types of controls:

(a) Technical Controls: This involves providing technical support or additional access to users to get them satisfied.
For cloud systems that hosts reconfigurable hardware or FPGA based platforms and provides the FPGAs to users in return of a license fees, it tells the users of how to use the services or tells them how to develop their custom bitstreams and upload them. These are done to prevent the user tasks from being stolen by other users or third party materials hosted in the cloud.
(b) Administrative Controls: This involves negotiating price with the users to making them satisfied for using a service and even giving access control.

Generally, the CSPs have mobile applications or a web interface, where they provide a graphical user interface (GUI) that lists the possible types of services that the user can use and what price they need to pay. They also have the provenience for the users to contact if they need any help.

(c) Physical Controls: This involves changing the system components like software or hardware products. This can take place with respect to time or when some components malfunction.

If any malfunction is detected by any of the agents, then alarm is raised and the hardware platform is changed. If anomaly is found in the FPGA fabric, then to ensure protection to user tasks, these are outsourced to another FPGA platform to ensure their completion before deadline. However, if malfunction is detected in any of the third party bitstreams, then the bitstream is replaced with another one and the vendor that supplied the malicious bitstream is blacklisted and all resources procured from it is prevented from usage in future.

3.1.4 Framing a Security Policy

For framing a security policy, there are a number of stages involved. This depends on the organisation, their objectives and what type of system platform is provided. The mechanism of security policy framing is depicted in Fig. 3.4.

This involves:

(a) Initial drafting: Discussions are performed with employers, employees and potential users to develop a system. Based on the system, the security architects develops related security policies and formulates them, thus, prepares a draft.

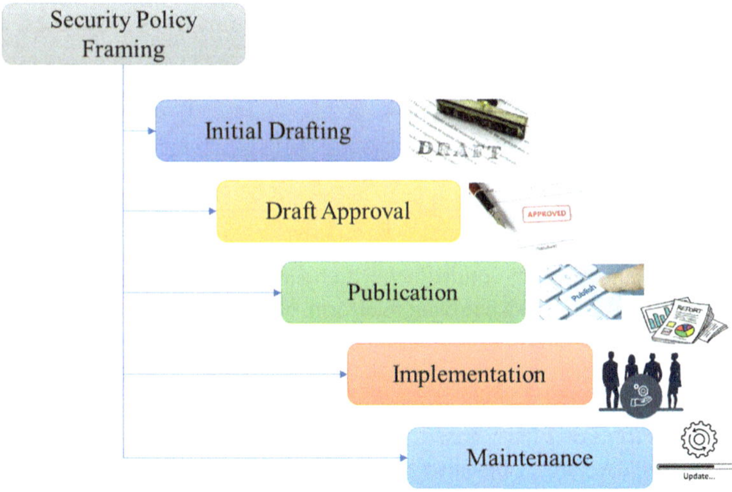

Fig. 3.4 Security policy framing

For example, when a CSP decides to host FPGAs in the cloud infrastructure, it holds meeting with its employees directly. For the users, it sends out questionnaires or potential feedback modules where the users can send their comments and feedback. Based on that, the security architects decides on what policies are needed.

(b) Draft Approval: The draft is checked by the employers and if satisfied, it is approved, else another draft is initiated.
For instance, based on the feedback given, it is suggested that the users should encrypt their files with data encryption standard (DES) algorithm that they want to execute in cloud FPGAs. However, during analysis it is found that this symmetric key encryption algorithm is not appropriate as this needs secret key sharing and storing of the secret keys for all the users. Hence, this draft may be denied. Another draft needs to be proposed in this scenario. For instance, the second draft is associated with digital signatures. This is mostly preferred. The method can be accepted and is approved.

(c) Publication: If approved by the employers, the draft is published and made available to its employees.
After approval of digital signature scheme, it is notified via the website or mobile application.

(d) Implementation: The employees and users act as per the strategies given in the security policy.
The users sends their files with digital signatures. The employees use the digital signature to verify that the files came from the appropriate users.

(e) Maintenance: The security policies are periodically checked and updated if required.

For instance, a new technology needs to be introduced. It may be associated with digital watermarking or use of physical unclonable functions. Then the security policies are updated accordingly, which involves the introduction mechanism of putting the digital watermarks or physical unclonable functions in the files, as well as the verifying mechanism.

3.1.5 Enforcing Security Policies

The enforced security policies must be based on several factors:

Strong authentication like multi factor authentication: For embedded platforms that contains a software and a hardware platform, multi authentication techniques are viable. Biometric information can be associated with softwares, while for hardware, physical unclonable function and hardware watermarking can serve the purpose.

Physical Security: For FPGA based platforms, physical security is associated not only for the FPGA fabric, but also must be ensured for the bitstreams or RIPs. Physical unclonable functions can be used for securing the FPGAs, while hardware watermarking is useful for ensuring security of the bitstreams.

Risk management: Hardware agents are generally associated with risk management. These basically works based on the observe-decide-act (ODA) paradigm. In parallel and distributed platforms and also in cloud and edge architectures, these can communicate with each other and outsource the task to ensure its completion.

Auditing: Auditing will be discussed in detail in Chap. 4. This is generally associated with record keeping. This enables to see which user used which FPGA platform and which bitstream for FPGA based infrastructures. This helps in backtracking to detect the malicious source.

Operational and Access Controls: These are essentially performed by the hardware agents deployed in the system architecture. These verifies the PUFs and watermarks before the resources are hosted or used by the system. These also checks the identity of the users before accepting any tasks to execute from them.

3.1.6 Value of a Security Policy

Other than enhancing the security and risk awareness of a system, a security policy also determines direct and indirect losses. While direct losses points towards financial losses that results from an incident, like loss of revenue. Indirect losses refers to other factors like loss of brand name or reputation of the service provider.

Hence, value of a security policy is associated to its contents, such that it can reduce and limit both direct and indirect losses and even limit the organization's liability to the losses that it incurs.

This is not only associated with the service providers, but the users or customers are also liable. Hence, the users and customers must be made aware of the security policies and they are aware of its contents and if they violate, they are liable. Even employees must be well adversed with the security policies and if they deter from its policies, they can associated with a penalty, if found accountable with a wrong doing. The absence of a security policy will give the freedom to both employees and users to exploit the system.

This is depicted in Fig. 3.5.

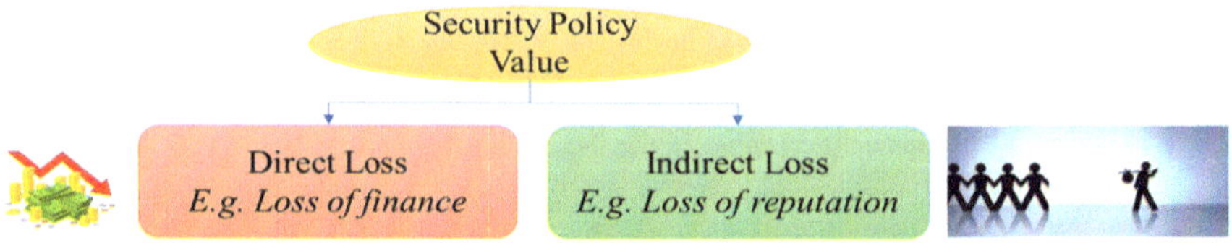

Fig. 3.5 Security policy value

3.1.7 Security Policy Depicting Corporate Culture

A security policy even portrays the culture of the service provider or its approach to business. For example, if it cannot provide a service, it can pay back the related fees to the customer.

The service provider may even give back the customer the related fees, if the customer is not satisfied with the service it provides. Replace the product it has sold, if found to be defective or is a counterfeit one.

This holds for cloud service providers (CSPs) that hosts FPGAs and bitstreams from different third parties. The CSPs needs to have security policy so that it can safeguard the interests of the third parties that are sending their reconfigurable intellectual properties (RIPs) or bitstreams to be hosted. CSPs need to put proper security measures to ensure that the secret data of the RIPs are not stolen by malicious users. If anything malicious occurs, the CSP needs to have a policy which can benefit the third parties, in such scenarios.

In addition to ensuring safeguarding the third party vendors, the CSPs also need to safeguard the users who are paying to execute their tasks.

Thus, the security policy of the CSPs must be made in such a manner that it will safeguard both the users and the third party vendors. It must also employ various detection techniques to identify who is responsible for a malfunction and put preventive measures.

3.2 Business Continuity Plan

A BCP essentially incorporates the mechanisms and strategies required to secure the mission critical services over a certain period of time Cerullo and Cerullo (2004). It includes multiple sub-elements like business impact analysis (BIA) and disaster recorvery plan (DRP), which we will discuss in this chapter. In general, a BIA is associated with the identification of critical functions and procedures of a buisness or service, while DRP portrays the restoration procedure for the critical system functionalities as quickly and as efficiently as possible, when some service disruption takes place.

The BCP lifecycle is depicted in Fig. 3.6.

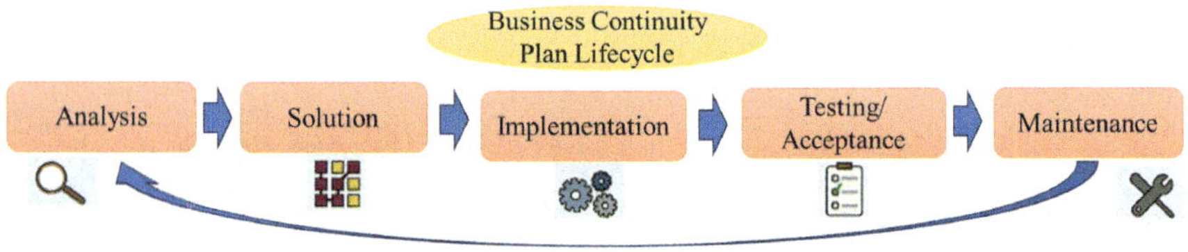

Fig. 3.6 Lifecycle of BCP

Service disruption can be either unintentional or intentional. Unintentional disruption can be associated with faults and equipment failure or due to environmental disasters, while intentional disruption is featured by introduction of trojans, use of counterfeited components, pirated components.

BCP is very essential for CSPs that hosts FPGAs and bitstreams, procured from third party vendors. It is quite common that the resources are associated with malware like hardware Trojans. Even physical resource components may be counterfeit in nature. Hence, the CSPs need a countinuity plan to ensure that abnormalities are tackled, if they occur at runtime. It is also important to identify the source of malicious resources and to blacklist them. This information must be shared to all so that no procurement of resources takes place from blacklisted vendors.

Moreover, such FPGA platforms are shared by multiple users. Some users may be malicious and try to disrupt the smooth occurring of other genuine users. Hence, these must be identified also.

3.2.1 *Steps of Developing a BCP*

Development of a BCP is associated with business impact, which is associated with identification of critical functions and hence, acts as a starting point to the development of the CSP. In addition to the BIA, it may have some backup policies and recovery strategies. Based on BIA, the recovery mechanisms for various scenarios are formed. After recovery strategies are standardised, test plans are formed to support the strategies. Finally, the authorities responsible for actualising the plans are trained appropriately. After proper training, they start implementing the plans. However, these are subjected to periodic reviews and updates. The steps for developing a BCP are depicted in Fig. 3.7.

This is particularly of interest for the CSPs that host FPGAs. This is because the resources are associated with execution of mixed critical tasks. High critical tasks may require more resources and time to complete their actions, while low critical tasks will require fewer resources and time. High critical tasks need high accuracy and their deadline is quite rigid, while these parameters may be relaxed for low critical tasks. Thus, based on the criticality, appropriate security policies must be defined and deployed. As the FPGA systems are associated with spatio-temporal scheduling policies, hence, the deployment of security needs to be changed with the task that will be executed. Moreover, CSPs may also take payments based on the criticality of task or even based on the security policies that will be deployed for it, hence, the impact needs to be understood from before and related mitigation strategies needs to be developed.

Thus, the fundamental steps for developing a BCP are:

- Understanding the business impact analysis
- Development of related recovery strategies
- Test the recovery plans

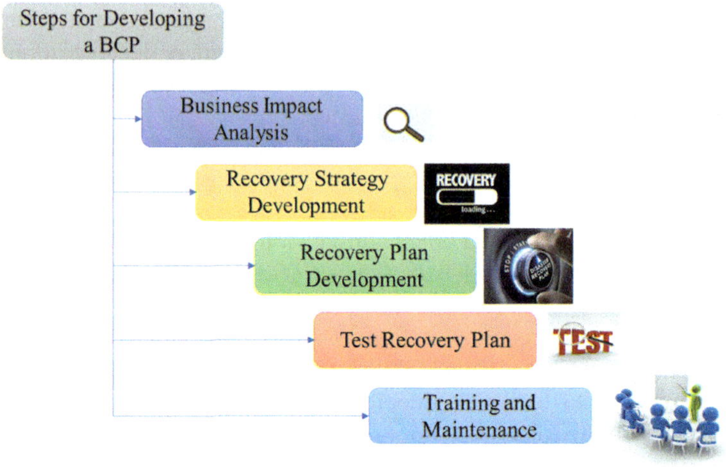

Fig. 3.7 Steps for developing a BCP

- Training and deployment
- Review and Update.

For instance, a high critical operation will need result accuracy and has a small time to execute. For this, a triple redundancy procedure for execution on resources from different sources and finally majority polling, is generated by the analysis team. This will provide a high level of security and definitly, the cost for deploying such type of security will also be high. However, if there is a relaxed scenario, then monitoring and checkpointing policy may be applied. In this, the online operations are continuously monitored and when a deflection from normal behaviour is observed, then roll backing to a previous checkpoint is used. This may cost lower. Thus, all the recovery plans are related to the criticality and these must be tested prior to deployment. For some high critical operations, backup mitigation plans may be needed. For this, while the first mechanism (triple redundancy with majority polling) can be the primary security policy. However, threat exists in nexus among the resource providers which may implant the same malware in all the resources supplied. In that case, the second policy (observing and rollbacking from checkpoint) can be used as the backup mitigation technique.

In some cases, automated hardware agents are used for monitoring and security. For these, different types of training algorithms need to be used and different datasets need to be deployed based on the related applications they are catering to. These agents needs to be tested, before actual deployment. Post deployment, they may need to be updated. So, it is better to perform the training and testing in software and finally, develop the bitstream for FPGA. If needs to be updated, the updated bitstream will reconfigure the related FPGA fabric.

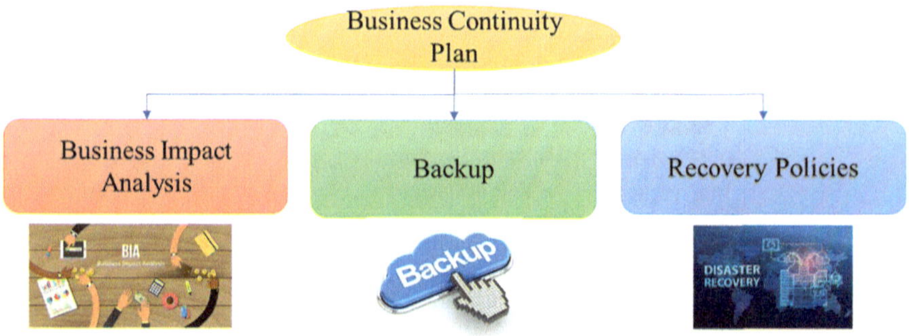

Fig. 3.8 Contents of a BCP

3.2.2 Business Impact Analysis

The three components of a BCP are business impact analysis (BIA), backup plans and recovery policies as depicted in Fig. 3.8.

BIA is an essential component of BCP, which is used to identify the impact to a service provider or an organization if any of its services malfunction. Such knowledge is significantly important, as it aids a service provider or an organization to identify what services are critical and devote their resources appropriately.

In addition to this, the BIA is also associated with evaluation of direct and indirect costs. While direct costs refer to immediate loss, which can be related to sales. Indirect loss refers to loss of reputation among the customers and long term business loss.

For instance, an organization will stop functioning if power outrage occurs. Now, the question lies whether this was caused intentionally or unintentionally. It is also important to understand whether the malfunction will affect critical operations or not. Such power outrage in a building will cause loss of server data and e-commerce websites, then it is definitely crucial and backup generators needs to be kept. However, if the power outrage occurs in a building that houses products for shipping and delivery, then no need to keep backup generators.

Similarly, if the CSP is hosting FPGAs that are associated with critical operations, then additional resources have to be kept, where outsourcing of task can be done if malfunction occurs or redundancy method should be used from the beginning. However, if the FPGAs are associated with non-critical task operations, then no such security measures are required to be deployed.

3.2.3 Maximum Acceptable Outrage

A significant portion of BIA is to determine the maximum acceptable outrage (MAO) for a service or a system that is offered by the service provider or an organization. This may also be termed as the maximum tolerable downtime (MTD). Recovery time

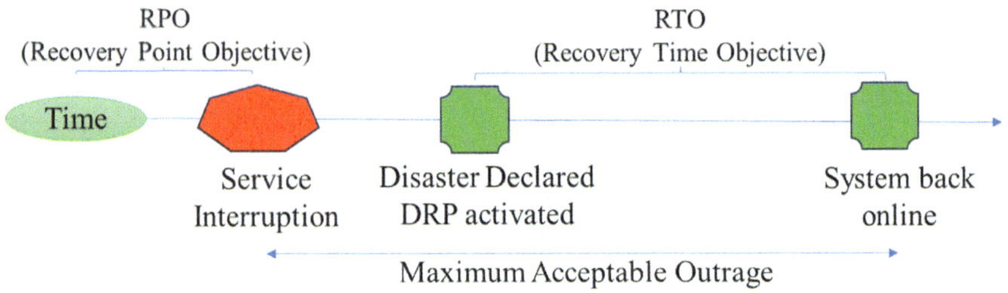

Fig. 3.9 Illustration of Maximum Allowable Outrage (MAO)

objective (RTO) can be defined as the maximum amount of time that can be provided to restore normality for a system, when it has faced some type of malfunction.

For instance, let's consider a building that stocks fruits. The fruits become stale if refrigeration is not available for more than 6 h. Then the MAO is 6 h. If the power outrage exceeds 6 h, then it can cause loss to the service. Hence, the service provider must make alternate arrangements within the MAO.

This is shown in Fig. 3.9.

3.2.4 Recovery Point Objective (RPO)

This is of particular interest for real time systems, as they are associated with deadline. CSPs that takes tasks from users have to complete it within deadline. For this, if any malfunction occurs, they need to be outsourced to another resource for execution. Moreover, on detection of vulnerability, the affected resources needs to be replaced, else the existing resources may not be able to handle the additional task burden for long. Thus, appropriate scheduling with security considerations have to be performed in the pre-deployment phase that needs to be adhered in the post-deployment phases.

During a malfunction, loss of data may take place. Recovery strategies must be put into operation such that the loss is minimal. A recovery point objective (RPO) is often associated with such situations that identifies the maximum amount of data that can be afforded to be lost.

Theoretically, RPO is zero, but practically, some data loss may be associated. Hence, redundant data storage is quite common, but this is associated with space. Thus, redundant data storage can be exhibited for high critical tasks.

For task schedules in FPGA based platforms, recovery points or checkpoints are quite common. This is done to re-execute the entire schedule when an error occurs. Such RPO based techniques aids in storage and cost reduction. Even acts as an energy efficient technique to develop a sustainable society.

This is shown in Fig. 3.10.

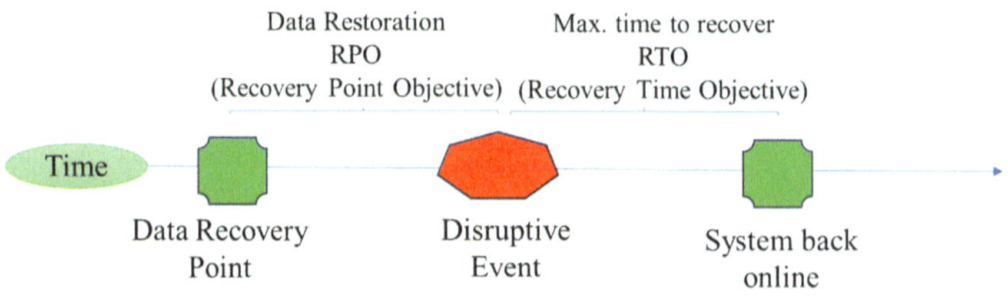

Fig. 3.10 Depiction of Recovery Point Objective (RPO)

3.2.5 BIA Output, Probability and Impact Analysis

A prime deliverable of the BIA is its ability to identify the financial loss that can result out of any malfunction. This is an important factor when we are associated with evaluating controls that can minimise the losses. As more resources and security measures need to be allocated for protection and completion of high critical tasks than low critical tasks. In systems that are associated with mixed critical tasks arriving from different users, BIA even prioritizes the tasks, as per their importance. Moreover, BIA also prioritizes the resources that are procured from third party vendors and hosted in the cloud platform. The resources that are of high quality are used for high critical tasks and accordingly.

Taking historical data into consideration is quite important in the development of BIA. Based on the historical data and its impact, the BIA can generate levels and organize its resources. It is important for a system to blacklist different vendors, whose resources are associated with a malfunction. The BIA needs to consider these. The BIA can develop a probability-impact graph, that can be used during scheduling and security at runtime, based on the criticality of an application.

This relationship is depicted in Fig. 3.11.

Fig. 3.11 Relationship between probability and impact

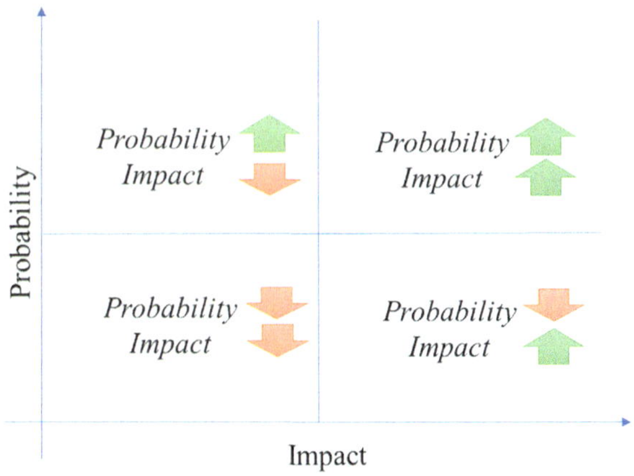

3.3 Disaster Recovery and Emergency Response Plans

A disaster recovery or emergency response plans aids a service provider to have some ready and routine solutions if some natural disaster or anomaly takes place. As these does not happen often, mostly they are not included in the initial stages of drafting. Moreover, it cannot be known in advance of what types of disaster or malfunction may take place with time and hence, drafting them in advance is quite a difficult action. Thus, the overall goal is to identify the various disasters or malfunctions that may take place and to fetch out some actions that can restore the system to normality.

3.3.1 Disaster Recovery Versus Fault Tolerance

It is important to note that disaster recovery and fault tolerance is not the same. While disaster recovery aids a service provider to recover its system after a critical malfunction occurs, fault tolerance aids the service provider to ensure that its system components continue to function after a malfunction or failure takes place.

For instance, CSPs that provides services with FPGAs may face a disaster if server location is associated with some environmental calamities that may affect the system resources. If some information is obtained in prior, then transferring the tasks to other server locations can be done.

However, some bitstreams that configures the FPGAs may function maliciously. Then, using alternate bitstream procured from another source and ensuring their timely completion is associated with fault tolerance.

3.3.2 Disaster Recovery Plan (DRP)

A DRP can be defined as a course of action that needs to be taken when some natural calamity occurs. This may be like some earthquake or floods that may destroy the servers, where the resources are present. Resources includes the FPGAs that are deployed in the cloud infrastructure. Unlike attacks, these are purely non-intentional in nature and beyond the control of human means.

Prior procrastination of the arrival of such disasters can be helpful in making a disaster plan. Disasters like earthquakes is difficult to detect early. However, disasters like floods can be procrastinated. A service provider may have a number of DRPs, based on the possible types of disasters that may take place in a region. Some DRPs may be associated with a line of actions over a certain time frame, possibly for scenarios like floods, while some DRPs may need quick actions ensuring protection and outsourcing of high critical tasks only. For instance, tornedos and floods are quite common in coastal areas like Florida, while not so common in Ohio.

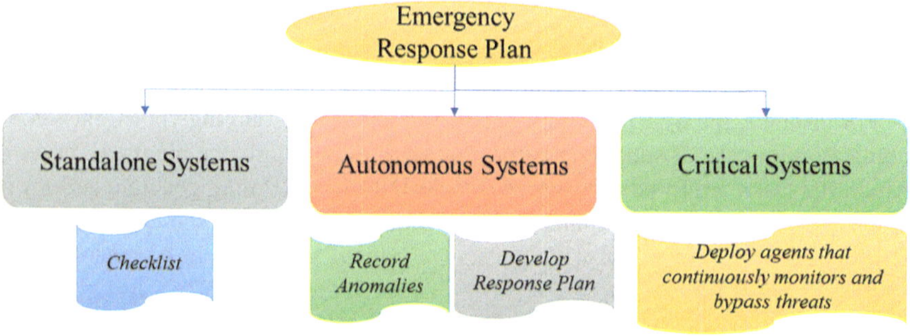

Fig. 3.12 ERP for different types of systems

3.3.3 Emergency Response Plan (ERP)

For reconfigurable hardware based systems, intentional and non-intentional anomalies or malfunction may cause the system to cease functionality or may even disrupt the functionalities of other users. Hence, disaster recovery is quite important. Moreover, many tasks are critical in nature and hence, emergency response needs to be provided. Thus, emergency response plan is quite important.

ERP for different types of systems is depicted in Fig. 3.12.

For standalone manual systems, a checklist is made for the person to follow, when an anomaly is detected. However, these are quite fine during experimentation and research.

However, for autonomous and headless systems, it is imperative that during testing phase, potential anomalies need to be recorded. Based on that, the controllers and hardware agents must be trained and programmed to act at runtime. If malicious actions take place, i.e. if cases occur that can lead to system malfunction or stoppage, then the controller and agents can move to an alternate course of action and ensure system normality.

Moreover, for critical tasks, additional system resources need to be deployed and emergency response and solutions must be adhered. Hence, making the appropriate plan in the pre-deployment phase is of utmost importance. Such a plan can be adhered by the controller or the hardware agents for facilitating recovery and prioritizing the execution of high critical tasks than low critical tasks.

3.3.4 BCP Versus DRP/ERP

With respect to service providers, DRP and ERP are a part of BCP. While the BCP has a wider scope and aids them to develop an action plan for a long term, the DRP and ERP is only associated with system restoration during an intentional or an unintentional system breakdown.

In general, the BCP illustrates the mechanism to keep critical functions running even during a disaster. However, a DRP or ERP has a narrower focus and is limited to recovery of the individual systems.

For instance, the BCP details the procedure of task outsourcing via wireless communication in case of sudden disaster. The BCP even details the alternate server in which the tasks are to be outsourced. While DRP or ERP may generate actions of how the resources can be replaced and in what time frame. In the extreme case, a DRP or ERP may detail out which tasks are critical and needs to be outsourced to complete within time.

3.4 Restoration Planning

A prime area of BCP is restoration planning. This involves:

(i) identifying the critical functions or tasks: This is mostly obtained from the BIA. For instance, if the CSP is taking more money for executing some tasks than others, then those tasks are much more critical. Moreover, in a composite task, the root task with several dependencies is highly critical.
(ii) Identifying restoration points or targets: Based on the MAO, potential resources are identified, where the tasks can be outsourced, so that they can complete before their deadline.
(iii) Plans for restoration: This step is associated with depicting the mechanism of transferring to other resources for execution. This is followed by identification of malicious resources, generation of alarms to change the resources and even blacklisting the vendors whose resources were associated with malfunction, so that their resources are never used in future to avoid any malfunction.

3.5 Drills and Exercises

A drill is associated with execution of response plans, in case of an emergency.

These are of two types: (i) Table top (ii) Functional.

Table top exercises are mainly based on discussions that are conducted in a closed room. This is related on how to train the system and how the various steps of the mechanism needs to be executed.

Functional exercise is associated with actual performance of the steps that are outlined in the plan. A false emergency is even created, and it is seen on how the system responds.

For the present case, the hardware agents that monitors the system activities are trained with various sets of data. During testing or in pre-deployment phase, it is seen how it behaves with malicious resources. It is seen whether it can generate an alarm or switch to other resources in order to meet the deadline of the real time tasks. Moreover, in post deployment phases, some buffer time is allocated to periodically test the activity of the agents with malicious resources.

Questions

1. Explain how a security policy can be used to ensure safety of multiple users in a cloud platform.
2. Explain the various phases of risk management and also state how they are useful when developing a cloud platform.
3. State how the technical controls, administrative controls and physical controls comes into action for hosting a cloud management platform with reconfigurable hardware.
4. Describe how a business continuity plan is useful for cloud systems. Also state the various steps in developing a BCP.
5. Describe how business impact analysis is useful for developing a connected system with reconfigurable hardware.
6. Explain with an example, the concepts of maximum acceptable outrage and recovery point objective for a server system.
7. Explain the difference between disaster recovery and fault tolerance for a cloud system with reconfigurable hardware.
8. State how a disaster recovery plan is helpful for cloud systems.
9. Explain how an emergency response plan is applicable to standalone systems, autonomous systems and critical systems.
10. State what is restoration planning and how it is useful for data servers.

References

Cerullo V, Cerullo MJ (2004) Business continuity planning: a comprehensive approach. Inf Syst Manag 21(3)

Fenz S, Heurix J, Neubauer T, Pechstein F (2014) Current challenges in information security risk management. Inf Manag & Comput Secur 22(5):410–430

Harris S (2010) CISSP all-in-one exam guide. McGraw-Hill, Inc

Chapter 4
Auditing

Objectives of The Chapter

- Understanding the 3 A's of Security (Authentication, Authorization, Auditing).
- Fundamentals of security audits.
- Audit Logs and User Accountability.
- Clipping Levels.
- Compliance Audit.
- Auditing Authentication Techniques.
- Configuration Management.
- Change Management.

4.1 Introduction

The AAAs of security:

(i) Authentication,
(ii) Authorization, and
(iii) Auditing.

This is depicted in Fig. 4.1.

K. Guha et al., *Cybersecurity for Reconfigurable Hardware Based Critical Infrastructures*,
https://doi.org/10.1007/978-3-031-67591-1_4

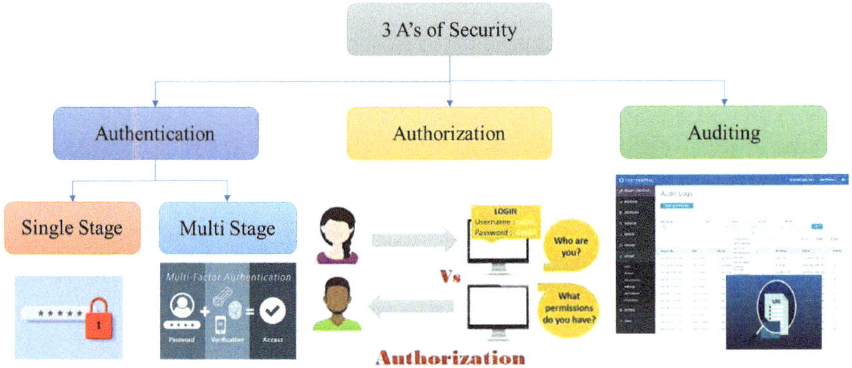

Fig. 4.1 Three A's of security

4.1.1 Authentication

Authentication can be defined as the process of proving or showing something is true or genuine.

In practice, authentication is carried out in various ways. These can be simple as password matching, while complex as biometric authentication. However, the present trend is multi factor authentication, where a combination of password and biometric verification is used. Such type of authentication is not only associated with testing of the memory capacity of an individual, but also depends on the individual's physical traits (password facilitates an individual's memory knowledge, while biometrics focusses on the individual's physical traits).

4.1.2 Authorization

After authentication is successful, the individual is granted access to use a system. This is called Authorization. Granting permission to unauthorized people can jeopardize the system, which may lead to system modification that may cause erroneous results, leakage of secret information and denial of service.

4.1.3 Auditing

Finally, auditing. Normally, auditing is associated with finance. However, in computer systems, auditing is associated with tracking the actions and records of an individual. For example, a person has a certain number of licenses to use and if he reaches the maximum limit, he is not granted access anymore. This is done via system audit. This phase generally occurs at runtime, or after the system is deployed.

Various types of audit logs exist on networking and computing systems. Regular logs are present in operating systems that record activities on the computer. Application logs generally record information of various applications at runtime. Network logs list the data of network devices like routers and firewalls. Activities of the proxy servers can be found in the proxy server logs, which allows an organization to track websites which their employers access via their web browsers and the time they spend over there. Even all activities associated with a firewall like the traffic which the firewall passes and which it blocked can be found in the firewall logs. Together, they aid to document the activities of different individuals accessing the system and networks. These facilitate administrators and auditors to track the different types of activities of the system.

An important phase of auditing is review, where the reviewers (or the auditors in the present case) look for anomalies in the logs. For instance, a long string of failed attempts to log in as an administrator in the system points towards a brute force attack.

Thus, when the three A's, or authentication, authorization and auditing are used together, then only authorized entities can use the system in proper manner.

4.1.4 The Three A's for Reconfigurable Hardware Based Infrastructures

Authentication for reconfigurable hardware or field programmable gate arrays (FPGAs) is associated with simple key based authentication to hardware watermarks Jai et al. (2003) and physical unclonable functions (PUFs) Guajardo et al. (2007). Key based authentication is associated with matching of a string of 0's and 1's. Hardware watermarks authenticate the bitstreams of reconfigurable intellectual properties (RIPs) that are essentially associated with secret functions implanted in the sea of HDL codes. PUF based authentication is associated with authentication of the FPGA fabric that involve use of secret functions, whose response is based on unique properties of the base semiconductor material of a device, which varies with the change of environmental factors.

For e-FPGA platforms, a software counterpart like verification via biometrics (for platforms that comprise of a processor) can be added with the traditional hardware authentication mechanisms to make it stronger.

For distributed and parallel systems featuring FPGAs, the agents need to execute one another's tasks. But before doing so, they must be certain that the task requests sent by fellow agents are original and not associated with malware. Hence, they authenticate each other based on traditional encryption-decryption techniques or unique delay-based signatures.

Remote authentication is favored for cloud and IoT platforms, where FPGAs are hosted and made available for users as a part of hardware as a service (HaaS), where user bitstreams are needed to execute Guha et al. (2019). For this, a certain region of the FPGA fabric and memory space is allocated to the user for choose a bitstream available in the repository of the cloud and execute it in its allocated memory space or executing its own bitstream. For the former case, the user needs to ascertain that the bitstream is original and not fake. Hence, they verify via the unique watermarks in the bitstream. Along with this, the user must be sure enough that the FPGA platform that the user is using is original and not hosted by any malicious entity, which is done via the PUF. For the later case, the cloud system must verify that the bitstream has originated genuinely from the user and not from any malicious source, which is also done via the watermarks inserted in the bitstreams.

After authentication, access is granted to users. For standalone platforms, users get the full access to the system to configure their bitstreams and execute their task. In some cases, the users may divide the FPGA platforms and execute multiple tasks. However, authentication is done only once. For e-FPGA platforms, the associated software may handle the task or can be done manually by the user, as per the system architecture. While for parallel and distributed systems, this is done by the related agents. However, for cloud and IoT platforms, where a single FPGA is partitioned into several virtual portions and each portion is allocated to a different user to execute their task, multiple authentication has to be done, each for a particular user. After individual authentication, access to the virtual portion of the FPGA fabric is given to the specific user to configure their bitstreams and execute their tasks in the related FPGA partition.

In this chapter, we will discuss the various audit mechanisms and how it is involved in cybersecurity not only for software platforms, but also for reconfigurable hardware platforms.

4.2 Security Audits

Security audits is associated with examining the practices and operations of an organization and determining whether those conform to the policies of the organization Rasheed (2014).

The benefits of security audits are:

(i) Verifying the quality of the current security strategies and whether it is adequate.
(ii) Checking that the training performed related to security are working.
(iii) Detecting any extraneous resources.
(iv) Reducing the cost by removing any unnecessary resources.
(v) Checking flaws that may have been introduced by the adoption of new technology or processes.
(vi) Proving that the organization is complying with all the rules and regulations set by the Government.

The organization can facilitate this via an internal audit or an external audit.

In internal audit, security auditors of the company monitor the existing practices of the organization and checks the security aspects. While in external audit, security auditors from outside comes to inspect the practises of the organisation. Both the internal and external audits provide a report on the organization's security policies and practises and checks whether they conform to the satisfactory standards.

Such internal and external audits are performed periodically like once a year. A lazy approach may be followed in some cases, where on the occurrence of any security events, the security audits are performed. However, when laws, regulations and guidelines are changed by the government and new security policies must be introduced, then the security audits have to be performed by default. Auditors can even check records of past audits to see what changes were made over the years and how they reflected and adhered to the standards. Even details of security incidents that occurred in the past are taken into consideration.

For reconfigurable hardware-based infrastructures, internal audit is mainly handled by agents that continuously monitor the operations of the system. Such agents may be either deployed in lightweight software processors or as individual hardware agents in distributed and parallel systems or in the cloud platforms. They continuously monitor the activities of the host FPGAs and maintain the logs. At regular intervals of time or in case of discrepancy, alarm is generated. Such cases are handled by manual interventions and thus, external audit comes in.

This is shown diagrammatically in Fig. 4.2.

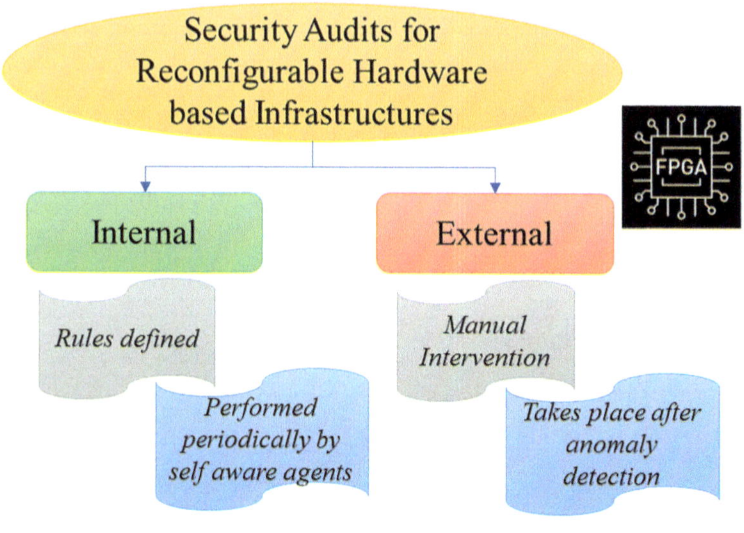

Fig. 4.2 Security audits for reconfigurable hardware based infrastructures

4.3 Audit Trails

An audit trail involves a set of log records occurring on a system or network via which events can be tracked. Based on these log records, auditors can re-create the events and determine what happened during the incident.

The main points that need to be considered while tracking via an audit trail are:

(i) Whether the audit trail generates a full user activity: This involves who, what, when, where.
(ii) Strength of the access control mechanism to the logs: If the access control mechanism is weak, then an intruder can easily delete log records.
(iii) Backup of log records in remote servers: If in any case, an intruder is able to bypass the security mechanism and deletes the log records, yet the tracking can be done based on periodic backups taken and stored in a remote server. Invasion to multiple places at the same time is practically not viable.

Log retention policy is used in many places where backup of the log records is kept for a substantial period of time like 3 years or 5 years, after which it is deleted automatically.

For instance, if an intruder targets a web server in a demilitarized zone (DMX), firewall logs can be used to identify the path of the intruder to invade the DMZ. We can even use the audit logs on the web server for tracking an intruder for breaches on the web server that involves firewall logs, system logs and security logs.

For hardware systems, such audit trails are particularly important to detect malicious users in multi-tenant systems. The tracking is done manually for the standalone platform. For e-FPGAs, the software aids in the tracking activities and maintains a log along with the log of software components in the internal memory of the SoC. The agent keeps the track for its host FPGA in a distributed platform. As resource is limited for e-FPGA platforms and distributed systems, hence, essential records are only kept. Finally for cloud platforms, as resource is not a constraint, hence, proper track records are kept as which user task accessed which memory blocks and for what amount of time.

Figure 4.3 shows the components of audit trails.

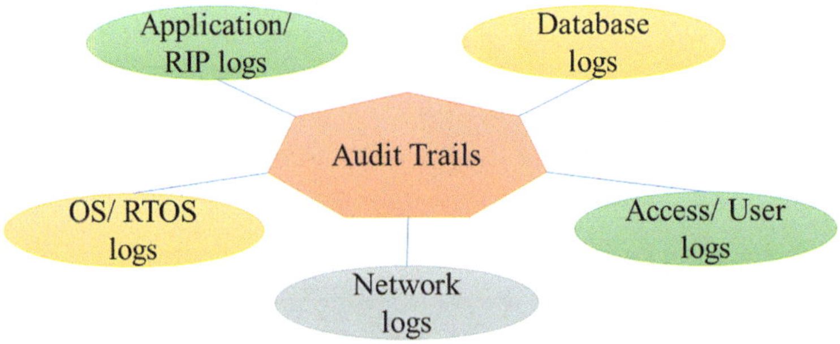

Fig. 4.3 Audit trails: a pictorial representation

4.4 Audit Mechanisms

The two strategies that are generally followed for auditing are.

4.4.1 Auditing Through Log Records

The audit logs record activities of access to different resources of the system and can be used to track the activities of users.

For general systems, the audit logs may be on the routers, firewalls, routers, etc.

For reconfigurable hardware-based platforms, the audit logs comprise which users are allocated which partition of the reconfigurable fabric. It even consists of the different bitstreams that are configured in the various partitions of the reconfigurable fabric and for what duration of time. It also consists of the memory access records of the users. These are stored in the local or centralized server.

Other than system access, logs are even kept for bitstreams that are procured from different third-party vendors. It records the time when a bitstream was included in the repositories of a server and when it was updated by the third-party vendor and its removal time.

Even system-based log records are kept, where the log records indicate the number of FPGAs that are deployed in a cloud or edge platform, from which vendors the FPGAs were procured and whether they were changed or updated with time.

4.4.2 Auditing Through Inspection

Security audits are performed at periodic intervals to check whether the resources of the organizations are being utilized as per their usage policy (even underutilization of resources will incur additional loss, while overutilization of resources will affect the assets). For instance, in generic systems, audits are made after periodic intervals of time to see if any resource requires maintenance or not.

For reconfigurable hardware-based infrastructures, audits are performed at periodic intervals of time to check

(i) the quality of the reconfigurable fabric
(ii) the number of times it was used and by whom during the period
(iii) whether it needs any replacement.

For real time scheduling on reconfigurable hardware infrastructures, the scheduler adheres to the various policies so that all the FPGA resources and their related virtual portions are equally utilized.

Audits are even carried out for employees of the organization to check whether they are adhering to the policies, rules and procedures framed by the organization.

4.5 Holding Users Accountable with Audit Logs

It is mandated by the National Institute of Standards and Technology (NIST) that federal agencies follow these essential guidelines:

- Generate, safeguard, and keep any records pertaining to audits of information systems that are necessary for the purpose of observing, analyzing, investigating, and reporting any behavior that is illegal, unreasonable, or improper; and
- Get everyone who uses the system to sign a document proving that they are responsible for their own actions.

With strong authorization mechanisms, the audit logs can be used by administrators, security personnels or auditors to hold a user accountable for any unauthorized or malicious operation.

For generic systems, an audit log entry depicts entry or starting access of a user like reading a file from the system.

For reconfigurable hardware system, granting access means providing the rights to a user to configure a FPGA or a part of it with its bitstreams. The scenario is more severe here as the user has direct access to the system hardware and hence, the log entry is quite important. This is because, in the event of malicious event, the user can be tracked based on the log entry.

An important factor here is authentication. If weak authentication is given like a password, then the user can deny his access or tell somebody else impersonated him. However, multifactor authentication like a combination of password and biometrics will make it difficult for a user to deny.

For FPGA based infrastructures, hardware watermarking for bitstreams and authentication for the FPGA fabric is done based on physical unclonable functions.

Reconstructing the events that leaded to a malicious incident is the primary purpose of auditing. Other than this, auditing can also serve as a deterrent. This simply means that users will not engage in unauthorized activities if they are aware that records are being kept of their activities.

For example, in reconfigurable systems, if records are kept about the memory access made by the users, then the users will refrain to access memory locations that are not allocated to them. Even users will not try to allocate their bitstreams in some other empty reconfigurable partition.

A common attack is the row hammer attack where users try to access its own memory rows in such a manner that it will cause bit flips in adjacent memory rows of other tenants. However, if log records are kept and in case of an abnormal activity, tracking the neighbours who possibly created the row hammer attacks can be done.

The components of audit logs is depicted in Fig. 4.4.

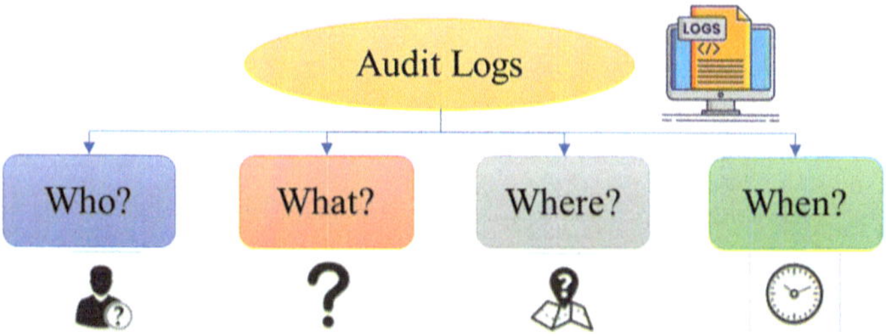

Fig. 4.4 Components of audit logs

4.6 Clipping Levels

Generating alarms is common for many systems. This occurs after a pre-defined number of events takes place, which is termed as the clipping level of the system.

Clipping levels is a key part in audit reduction tasks. Based on clipping levels, unnecessary log records are removed. Automated audit reduction tools examines a log and retrieves relevant data from the log based on the clipping level and ignores irrelevant data. Thus, the clipping level ignores normal user errors, till it reaches the pre-determined threshold.

For software systems, a classic example of the use of clipping levels is a report that is based on failed log-on attempts. This clipping level is used to generate a report for failed log-on attempts in a given period, rather than reporting for every single event.

In FPGA based infrastructures, a user may execute its tasks in different virtual portions of different FPGAs. Even a vendor may provide several bitstreams that are stored in a repository. If malfunctions are found with tasks associated to a particular user or on bitstreams procured from a specific vendor, then it is important to stop activities of the user and prevent selection and execution of the bitstreams procured from that vendor. Stopping user activities or preventing bitstream utilization is not started until a specific number of instances has been reached, which is defined as the clipping level. The agents which monitor the activities of user programs may suspend the user program or block the user from further access in any of the resources and even stops usage of bitstreams procured from third party vendors in such scenarios.

Clipping levels for reconfigurable hardware based infrastructures is shown in Fig. 4.5.

4.7 Compliance Audit

Along with security audits, there are compliance audits.

Compliant audits are a type of security audits that facilitate organizations to show that they are following (i) rules (ii) standards (iii) governance, (iv) regulations (v) transparency (vi) policies (vii) law (viii) requirements.

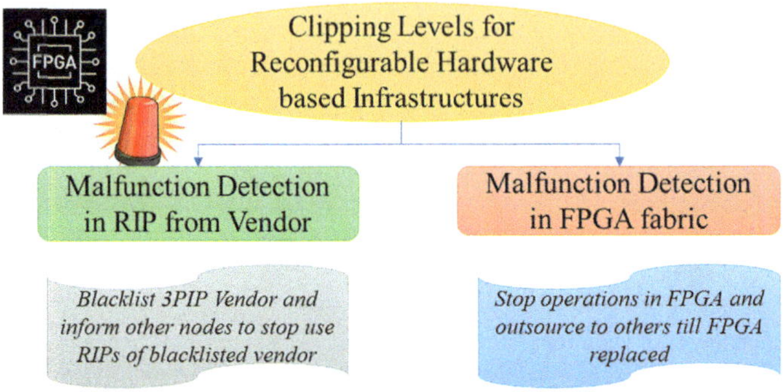

Fig. 4.5 Clipping levels for reconfigurable hardware based infrastructures

The internal compliance audits give proof to outside third-party entities that the organization is complying with the government rules and strategies.

For instance, US companies that are associated with development of health related equipment's and products or handles health related data, they must comply with Health Insurance Portability and Accountability Act (HIPPA).

Presently a globalization strategy is followed in product design. In case a security incident occurs, those companies which comply to the standards are more safe than the companies that did not comply with the standards. During inspection, products that did not meet the compliance standard are to be inspected first. Moreover, if after an audit, it is found that an organization has not affirmed to the compliance standards, then they can incur heavy fines and penalties.

Similarly for FPGA based systems, cloud service providers that host FPGAs and related bitstreams are to provide a set of rules and standards, based on which the vendors must design the bitstream and the FPGAs. The reason being several cloud service providers have their internal security configurations that protects their internal infrastructure. Such guidelines will provide a compliance related to the service products.

Figure 4.6 shows the key components of compliance audit.

4.8 Auditing Authentication Techniques

4.8.1 Software

Passwords: checking whether strong passwords are used, and it complies to the standards (like it must have an uppercase letter, a lower-case letter, a special character and a number and it should be at least of 8 digits). For this, a password audit, a password cracking tool is to be generated that can see how quickly the passwords can be broken.

Biometrics: must be re-loaded after a certain time frame as the characteristics of a person changes after a significant amount of time.

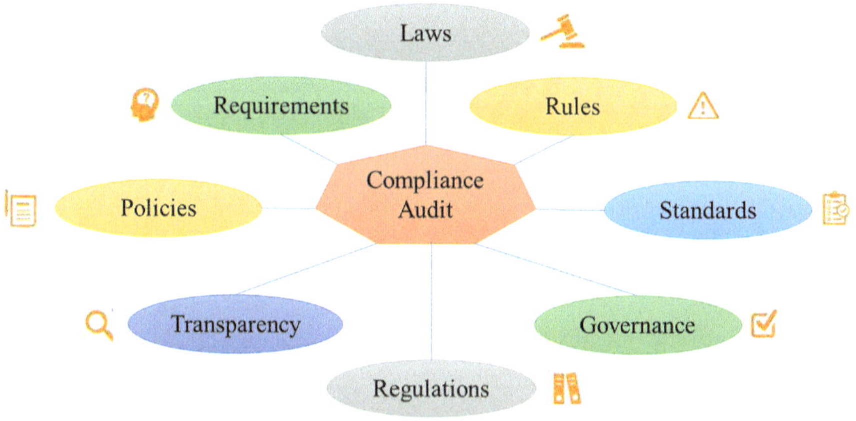

Fig. 4.6 Key components of compliance audit

Like facial passwords to be re-taken after around 10 years, voice-based passwords must be updated after around 5 years.

4.8.2 Hardware

A series of challenge responses pairs (CRPs) are maintained for the PUFs and watermarks. Auditing in this scenario refers to the type of PUF or watermark that is to be generated, which involves the number of CRPs that can be generated and how varied they can be. This also involves how varied the CRPs can be if the PUF functionalities are deployed in different regions of the hardware fabric.

4.9 Configuration Management

4.9.1 Overview of Configuration Management

The practice of ascertaining that systems are configured in a correct manner, or the configuration of the system is performed with adequate security is commonly termed as configuration management.

The practice of configuration management entitles an organization to configure its systems with a secure baseline and ensures that the system maintains a configuration that is free from vulnerabilities.

In general, organizations do not want to alter the configurations of their systems unless some security incidents occur or they need to update their configurations based on the updated rules and regulations of the government. For instance, administrators want similar system configurations on their servers when they start.

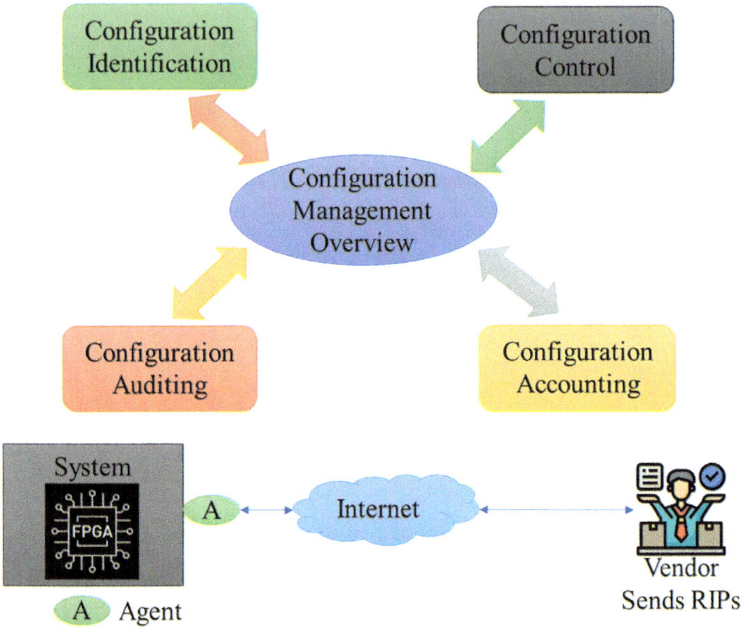

Fig. 4.7 Pictorial representation of configuration management

For reconfigurable hardware based platforms, this is of particular interest, as improper configuration of the FPGA fabric by bitstreams will lead to severe anomaly. Moreover, for spatio temporal scheduling aspects, it is not only important to note which bitstream is configuring which partition of an FPGA, but it is also quite important to note the order in which the bitstreams configure a particular portion of the FPGA fabric.

A pictorial representation for configuration management is depicted in Fig. 4.7.

4.9.2 The Cycle of Configuration Management Include

Configuration Identification: When a system encounters a security gap, the system configuration is reviewed and vulnerabilities are tried to find. The target is to identify a configuration that can mitigate the security gap.

For FPGA platforms, it is important to identify which bitstreams coming from which users are to configure which virtual portion of the FPGA fabric and at what instance of time.

Configuration Control: After a successful configuration is identified and the system is configured, the control settings of the new system configuration need to be carefully managed and ensure that it is safe from different security attacks.

Hardware security protocols manage and take care of the system after it is configured. Identifying the source of the tasks before executing them is the key, which is based on hardware watermarking.

Configuration Accounting: This stage relates to the different configuration accounts that can be possibly present in a certain period of time. Keeping a record of which configuration account configured the system at what time frame is also important. Managing the different configuration accounts is even necessary.

This is generally done by the controller or the agent which controls and organizes the order in which bitstreams are collected from the repository and dumped into the FPGA fabric.

Configuration Auditing: This stage is associated with periodic reviews of the configuration settings of the system.

This action is again performed by the agents, which continually observes and monitors the operations associated with the host platform.

To ensure security, administrators take the aid of various hardening techniques like removing or disabling several services that are no longer used. Even changing the default passwords of the servers is a common practice. Even changing or modifying the configuration of different groups of servers to meet the needs for the groups is also a common practise. For example, administrators group the web servers and the file servers separately and use different protocols for configuring the web servers and use different protocols for the file servers.

Information technology system accounting standards, including configuration control, are laid out in Federal Information Processing Standards Publication 200 (FIPS Pub 200). According to it.

- during the development life cycles of each system, create and update baseline configurations and inventories of the organization's information systems (including software, firmware, hardware, and documentation); and
- determine and enforce security configuration settings for IT products used in information systems within the organization.

It is important to follow these guidelines, even though they are not unique to systems that use FPGAs.

4.9.3 Imaging for Configuration Management

For general systems, a reference computer is chosen as the starting point, which is generally the server. In the server, all the applications are installed, and the operating system is configured based on the needs of the organization. For example, configuration management can be done for auditing policy, password policy, account lockout policy, etc. Such configuration is set as an image file. The image is tested, and administrators capture it using an imaging software. This image can be deployed to other systems and the systems configured with similar image can boot and start up with similar configuration settings as the reference computer.

For reconfigurable hardware based systems, the central controller which contains the reconfigurable hardware unit needs to be configured by various reconfigurable intellectual properties (RIPs). These are associated with different functional operations and performs different functionalities. For auditing in this case, generic PUFs and watermarks are deployed.

4.9.4 Group Policy for Configuration Management

In some networks, as in Microsoft networks, a Group Policy is used for configuring the systems, which facilitates a centralized administrative mechanism for all the systems. In this, an administrator configures a setting with a Group Policy object like opening a firewall port, which applies to all the computers that adheres the Group Policy. This change is associated to all the computers in the system, whose number can be any, from 50 to 5000.

In distributed reconfiguration based systems, the agents communicate with each other and makes related decisions. For instance, if some vendor is blacklisted by a particular agent, this information is sent to all other agents and they too blacklist the particular vendor. Moreover, if some resource stops working, the agents offloads related tasks or workloads to other resources via the agents.

4.10 Change Management

When systems are interconnected, then a change in one may result in unintended changes in another. Hence, a change management policy is of utmost importance in systems.

For instance, in software network systems, we can consider that a server is hosting an application, which opens TCP port 4567 in a server. This event may be considered faulty and anomalous by some other system administrator and may close it. This action will break the application running in the server. Users associated with that operation will be blocked of the services and its recovery time may take a long time. This is of particular importance for distributed reconfigurable systems. This is because may works are spread across a network in different resources. If anyone of the resource is associated with a fault, then synchronization of the entire system will be affected. All users will be affected.

So, a control change mechanism is needed. Even such a mechanism aids to create a documentation trail. If any malfunction occurs or a catastrophic failure result at any time in future, such a documentation trail will aid the administrators to restore it to its last working state.

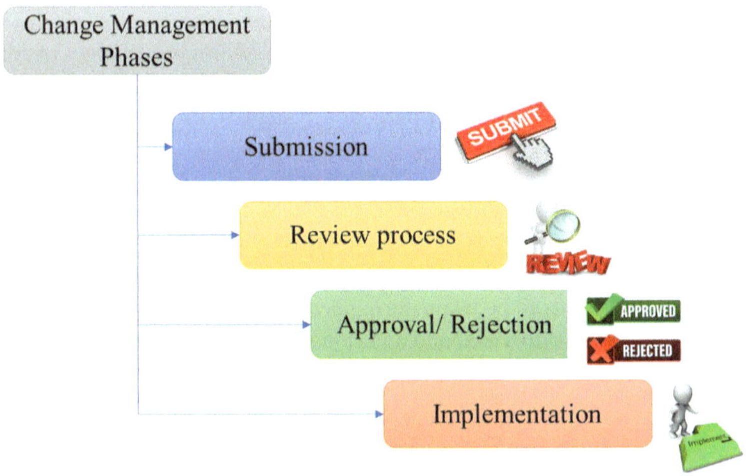

Fig. 4.8 Change management phases

4.10.1 Phases of Change Management

For a change to take place, the stakeholders first propose what changes are needed in the system. Such a proposal is examined and reviewed by multiple entities, before taking the final decision of whether to proceed with the change or abandon it.

The generic phases of change management is depicted in Fig. 4.8.

The various phases are:

4.10.1.1 Change Submitted

This phase is associated with observing the operations and finding the need of a change. This proposal for having the change is submitted, including the current system configuration and future system configuration, along with justification.

For instance, if the FPGA in a cloud platform requires a change. This can be either due to overload of tasks or may be due to an anomaly detected. For this, the required change is submitted to the decision making council of the organisation, which in the present case is the cloud service provider.

4.10.1.2 Change Reviewed

The proposal for the change is reviewed as discussed previously by multiple entities like the system administrators and even the top users. Such a process can be either formal or informal. Even a group meeting can be held.

The proposal is reviewed by a set of experts. The experts tries to see if the change in FPGA can handle the required task load or some additional changes needs to be made to ensure smooth working of the entire system as a change in one resource may hamper the synchronization of other parts of a system.

4.10.1.3 Change Accepted/Rejected

After the review, a decision is taken. If approved, then it is also analysed, whether any additional care needs to be taken during performing the change. If not, then, the reasons for rejection are provided and additional needs and possible alternatives are also given out.

Based on the expert decisions, the change is accepted or rejected by the decision making council of the organization, which in the present case is the cloud service provider.

4.10.1.4 Change Implemented

If approved, the change is implemented. For this, a series of stages is again followed. After the change is successfully carried out, a new documentation is made.

If the change is approved, the change of FPGA is made in the cloud platform.

4.10.2 Change Implementation

The various stages of change implementation are shown in Fig. 4.9 and also described as follows.

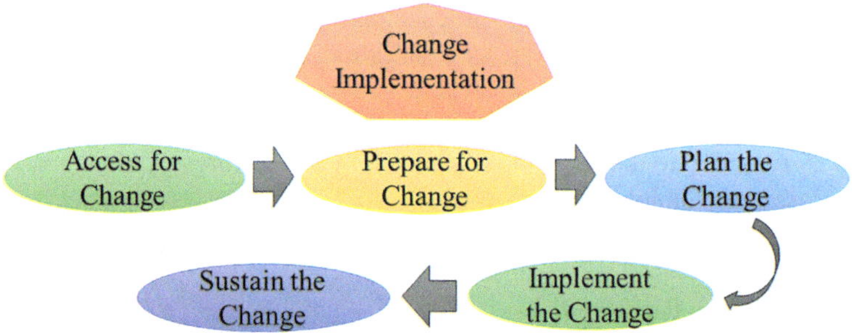

Fig. 4.9 Change implementation

4.10.2.1 Assess for Change

This is associated with analysing all the system components, as well as the connected components, whether they are ready to handle a change.

For instance, if malfunctions are found with RIPs from a particular vendor by a hardware agent in a distributed IoT environment. Then the agent blacklists the vendor who is supplying the RIP.

4.10.2.2 Prepare for Change

Notifications is sent to all the nodes of a system, as well as to all the connected systems, so that they can prepare the systems for the change. For instance, shutting down some active applications that may be hampered when the change is going on.

The hardware agent communicates this information to other nodes in the distributed environment. This it does by communicating with other hardware agents either by message passing or via writing in the shared memories.

4.10.2.3 Plan for Change

Appropriate planning to be done to feature the change. If the change is associated with multiple system components, then plan has to be done, which component is to be configured first and which next.

The change is to prevent usage of RIPs supplied by the blacklisted vendor and use RIPs procured from other vendors. It may also be the case, where a new vendor have to be contacted.

4.10.2.4 Implement the Change

The actual configuration changes are performed.

All the hardware agents implements the change, i.e. stops using RIPs sent from the blacklisted vendor.

4.10.2.5 Sustaining the Change

It is checked whether the change is suitable to all the system components and the connected systems. If any unintended consequences occurred due to the change, the changed configuration has to be rolled back and the process must be restarted.

It may happen that a previous operation was associated with a RIP procured from a blacklisted vendor. This may cause ripple effects in future. To prevent this, the hardware agents can perform redundancy based majority polling operations in free instances of different resources and be sure that no malfunction takes place.

Questions

1. Explain how Authentication is applicable for reconfigurable hardware based infrastructures.
2. State how authorization is useful for ensuring security of reconfigurable hardware based infrastructures.
3. Justify why auditing is recommended for ensuring security of applications associated with reconfigurable hardware based infrastructures.
4. Discuss how security audits can be performed for reconfigurable hardware based infrastructures.
5. State the different audit mechanisms and how an user can be held accountable based on audit logs.
6. Explain how clipping levels is useful for reconfigurable hardware based infrastructures.
7. State the key components of compliance audit and also state the auditing authentication techniques that are commonly used in practice.
8. Justify the importance of group policy in configuration management and how this can be applied to distributed platforms.
9. State the various phases of change management and how they can be associated with reconfigurable hardware based platforms.

References

Guajardo J, Kumar SS, Schrijen GJ, Tuyls P (2007) FPGA intrinsic PUFs and their use for IP protection. In: Cryptographic hardware and embedded systems-CHES 2007: 9th international workshop, Vienna, Austria, September 10–13, 2007. Proceedings 9. Springer, pp 63–80

Guha K, Saha D, Chakrabarti A (2019) Zero knowledge authentication for reuse of IPs in reconfigurable platforms. In: TENCON 2019-2019 IEEE region 10 conference (TENCON). IEEE, pp 2040–2045

Jain AK, Yuan L, Pari PR, Qu G (2003) Zero overhead watermarking technique for FPGA designs. In: Proceedings of the 13th ACM great lakes symposium on VLSI, pp 147–152

Rasheed H (2014) Data and infrastructure security auditing in cloud computing environments. Int J Inf Manag 34(3):364–368

Chapter 5
Monitoring and Analysis

Objectives of The Chapter

- Understanding Agent based Systems.
- Introduce Intrusion Detection Systems (IDS).
- Introduce Intrusion Prevention Systems (IPS).
- Knowing IDS Alerts.
- Understanding Detection and Analysis Techniques.
- Knowing Whitelisting and Blacklisting Techniques.
- Result Analysis for Security.

5.1 Introduction

For any type of systems, monitoring is an important aspect that aids the system to secure it at runtime from events that can potentially jeopardize their activities in future and lead to fatal consequences. Monitoring aids to detect and prevent malfunctions and also bypass future incidents Gibson (2011).

Based on the parameters that are monitored, analysis is carried out to understand whether the monitored event will lead to undesired consequences. If not, then it is best to carry on with the scheduled actions, else some bypass strategy needs to be sought or some remedy needs to be provided. A pictorial representation is shown in Fig. 5.1.

For reconfigurable hardware based architectures, these are generally performed by the hardware agents. Such hardware agents are mostly external entities. If they are made a part of the procured resources from third party vendors, then chances are present that they may also malfunction, when the resources malfunction. Such agents

K. Guha et al., *Cybersecurity for Reconfigurable Hardware Based Critical Infrastructures*,
https://doi.org/10.1007/978-3-031-67591-1_5

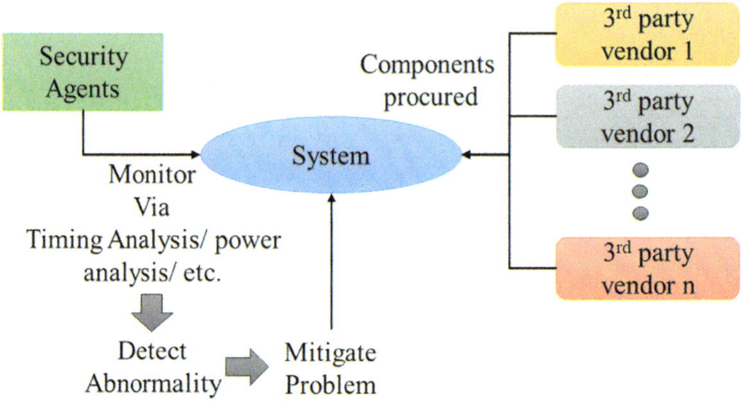

Fig. 5.1 Pictorial representation of agent based monitoring

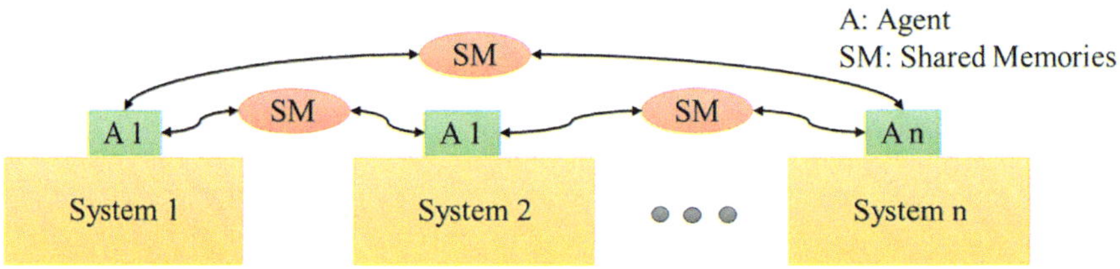

Fig. 5.2 Multi agent system

are generally of low overhead and has the ability to communicate among themselves via message passing and shared memories, as shown diagrammatically in Fig. 5.2.

Based on these, the concepts of intrusion detection system and intrusion prevention system is developed.

Intrusion Detection Systems Intrusion detection systems (IDS) facilitates continuous monitoring of events or activities of a system and its associated network and reports the events that may be associated with deviation from normal scenario.

Intrusion Prevention System Intrusion prevention systems (IPS) has more actions than an IDS. While an IDS is limited to detection of events and reporting them, IPS is also associated with remedial actions and bypass solutions that may prevent a system from a possible malfunction.

5.2 IDS

In general, IDS is associated with monitoring of different events Liao et al. (2013). For generic systems, these are associated with monitoring the networks and preventing them from intruders. For e-FPGA systems, such events may be associated with timing

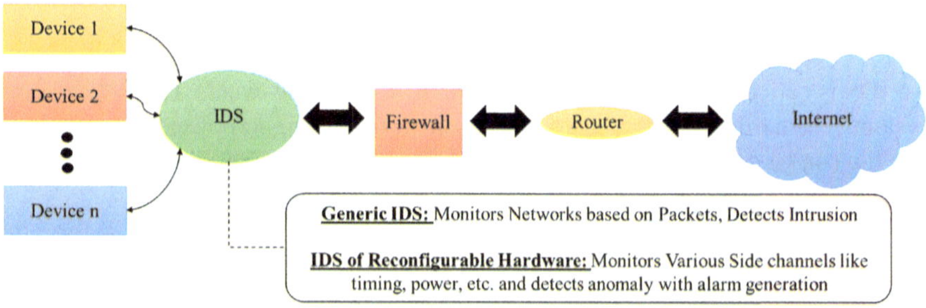

Fig. 5.3 Intrusion detection systems

activity for real time sub-tasks, power analysis, leakage current analysis, etc. The mail goal is to detect anomaly and stop the attack.

This is shown in Fig. 5.3.

5.2.1 Active and Passive IDS

IDA may be either active or passive in nature. The difference between active and passive IDS is depicted in Fig. 5.4 and also enumerated below.

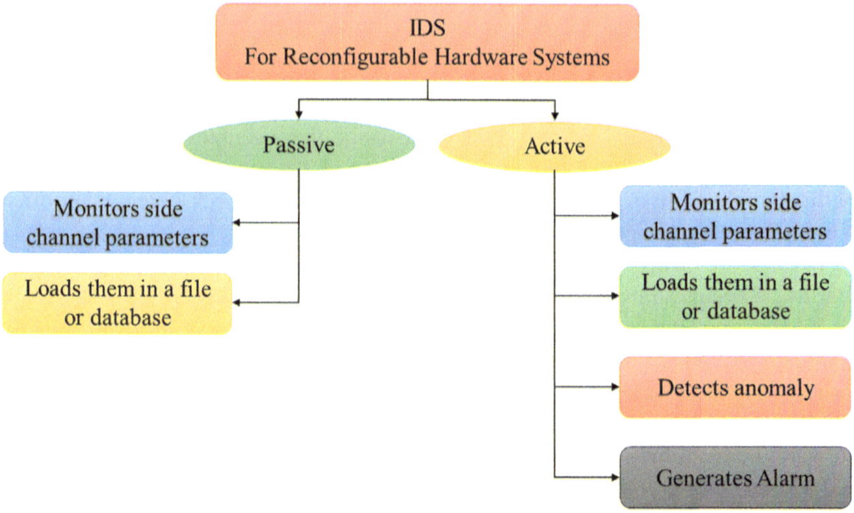

Fig. 5.4 Active and passive IDS for reconfigurable hardware based systems

5.2.1.1 Passive IDS

An IDS is called passive when it monitors and records the events. For FPGA based systems, the hardware agents that only monitors the timing or power and loads them in a file can be considered as components of a passive IDS. Based on the records of the file, future learning, prediction and analysis is carried out.

5.2.1.2 Active IDS

However, an active IDS may take some additional steps after detecting the malicious event. Like a passive IDS, an active IDS also monitors and records the events. In addition to this, it may also generate alerts, when a deviation from normality is seen. Hardware agents that generates alarms for replacing of system resources can be considered components of an active IDS.

5.3 Network and Host Based IDS

IDS can also be categorised as network based or host based.

5.3.1 Network Based IDS

A network based IDS (NIDS) observes and monitors traffic or network packets that passes over a network. Such NIDS are generally placed on various network segments like before a demilitarized zone (DMZ), or inside the resources of a DMZ and on internal routers. In this system, several nodes and agents are planted around the network in the form of routers, firewalls, and network appliances that together offers a unified threat management (UTM).

FPGAs are being widely deployed in routers and other network connectors. Hardware agents analyses the passing traffic over these resources.

Such agents can communicate with each other and tries to collect information about the system resources. For NIDS, the system designer installs the agents on various servers, routers, firewalls or on any device that has in the right place in the network to monitor the traffic entering and outgoing into the system. The NIDS can forward the recorded events to the central controller for furthur analysis and action.

This is depicted in Fig. 5.5.

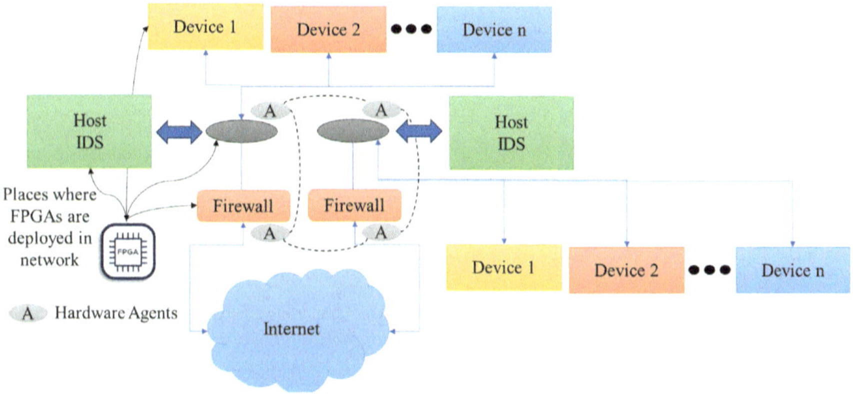

Fig. 5.5 Network based intrusion detection systems

5.3.2 Host Based IDS

Host based IDS (HIDS) monitors the events over a particular resource. However, there are limitations related to HIDS. Firstly, the latency. With the working of the HIDS, additional latency or time is associated which is not good for hard real time systems, as they have strict timing requirements. Moreover, installing HIDS in each of the systems is quite expensive. Thirdly, additional area and power overhead in the system is associated, due to the additional security agents. Hence, one HIDS agent is associated with a set of resources, where overhead limitations are present related to area and power. Hardware agents are associated with each FPGA resource in normal scenarios, while these are associated with a group of FPGAs for resource constrained environments. The number in the set is determined based on the maximum permissible area and power overhead.

HIDS is mainly associated with monitoring the activities of its host system and the operations associated with it. However, it cannot monitor the operations of the other resources and hence, needs to communicate with other agents to understand the behaviour of the entire network.

This is depicted in Fig. 5.6.

The best policy is to mix and match. While HIDS are associated with groups of resources, NIDS is used in intermediate positions of the network that connects the groups. Thus, the NIDS not only concentrates on monitoring the network traffic but also are useful for communicating between different HIDS that contains information about the individual or set of resources that it is monitoring.

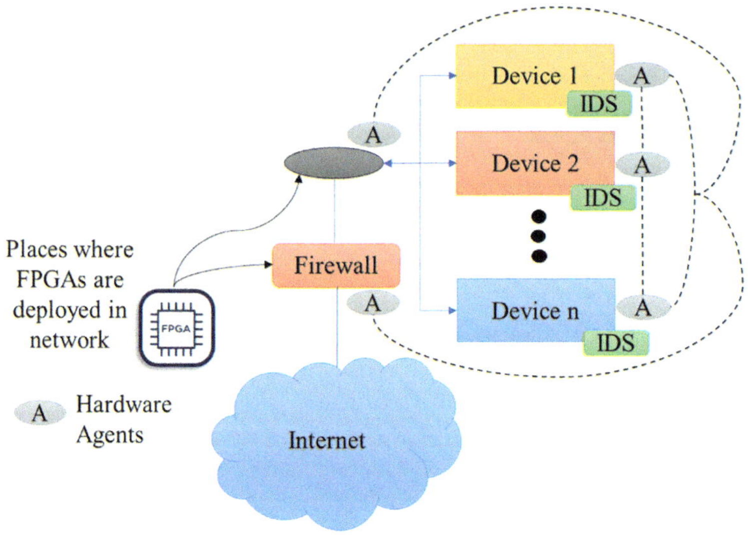

Fig. 5.6 Host based intrusion detection systems

5.4 IDS Alerts

On detection of an activity that is not normal, IDS informs the controller. This can be in the form of alert generation.

For this, on detecting any unnatural activity, the IDS generates the report and sends it in the form of e-mail or text message for generic systems. Even graphical alerts are common for huge systems associated with critical infrastructures like weather monitoring.

For resource constrained environments, this alert can be in the form of setting up a register from '0' to '1' or the glow of an LED in SoC platforms.

However, the noteworthy thing is that it specifies the point where vulnerability is detected. An alert does not mean that an intrusion is underway or a threat to a system has occurred. It simply means that a deviation from normal activity is encountered. More analysis is needed at that point to decipher the cause of deviation from normality.

It is quite common for IDS systems to give false positives, i.e. no attack is present, yet an alarm is raised. This may simply take place due to a change in environment like aging of system resources. However, such an alert is quite significant as it tells the system administrators that a change of system resources is needed in near future or system malfunction cannot be eradicated.

5.5 Intrusion Prevention Systems

An intrution prevention system (IPS) has some additional functionalities to IDS Chakraborty (2013). Along with detection of abnormality, it tries to decipher the cause and prevent it from occurring.

An IPS designed fully or partially with FPGAs or reconfigurable hardware performs much more than a normal IPS. Along with observing network packets for generic systems, the IPS is also designated to check the watermarks or PUF signatures for the incoming reconfigurable hardware IPs. Additional detection via side channel analysis can be carried out like IDS. If a mismatch of signature is found, it may blacklist the source and stop receiving any other materials from the blacklisted source. Generic structure of an IPS is depicted in Fig. 5.7.

Like IDS, IPS has also several forms and placed in different parts of a system. Like an IDS, an IPS can also be network based or host based.

5.5.1 Network Based IPS

In general, a network based IPS will be placed with a device like firewall that encounters the system traffic. For instance, at the boundary between a local network and the internet, such that it can monitor all the incoming and outgoing traffic and block malicious traffic. While the firewall is associated with protection of the local network from malicious traffic, the IPS aids to protect the network from attacks that can bypass the firewall. An IPS has the capability to modify the environment to ensure prevention of the entire system. For instance, the IPS can modify the ACL on a router if a pot scan is detected from a specific IP address. So, we can say that while an IDS may be either active or passive, all IPS are active.

We can also take another example where a server has been attacked with a SYN flood. In this, the attacker disrupts the transmission control protocol (TCP)'s three way handshake process and leaves half open connections on a server. Then the

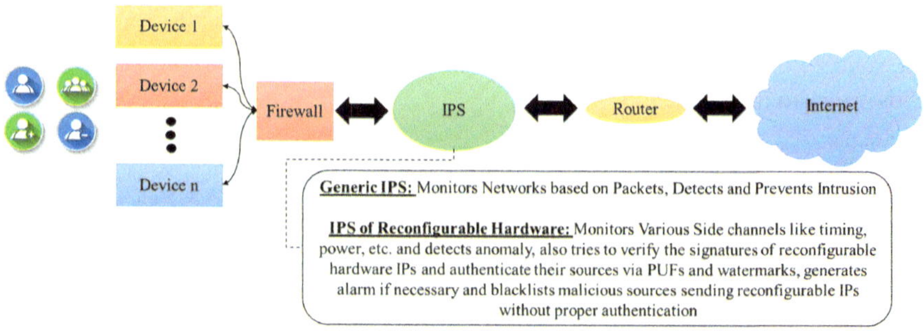

Fig. 5.7 Generic structure of intrusion prevention systems

Fig. 5.8 Working of intrusion prevention systems during SYN flood

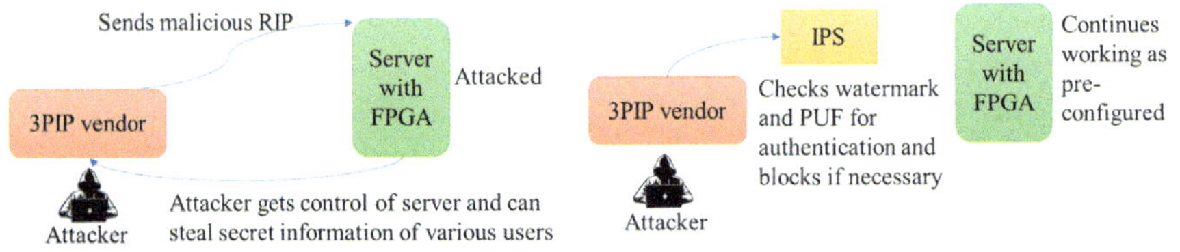

Fig. 5.9 Working of IPS for FPGA based servers

attacker sends the first SYN packet, which the server responds with the SYN/ACK packet. After this, the server waits for the ACK packet, but the attacker doesn't reply with the ACK packet. Thus, the server can have a number of half open connections that waits for the third packet in each case. The hardware agent which is associated with the working of IPS can understand the environment or detect such type of attack and tries to prevent this by blocking anymore traffic from the attacker. This can be done by modifying the ACL on the router or by closing the half open connections to free the resources on the server that are being consumed by the attack. This scenario is diagrammatically depicted in Fig. 5.8.

Present day servers are associated with several reconfigurable hardware or FPGAs. However, they are configured with reconfigurable intellectual properties (RIPs) that are sent from distant sources. Adversaries may intercept such RIPs during transit and send malicious RIPs, which if configures the FPGAs, will take control of the FPGAs and may steal secret information from the servers and send them to the adversaries or attackers. IPS systems will monitor them and check the watermarks and PUFs associated with them. If the signatures do not match, then these are blocked and also the sources from which these originates may be blocked and blacklisted in future. This scenario is depicted in Fig. 5.9.

5.5.2 *Host Based IPS*

These IPS are associated with each system resource or a set of system resources for a resource constrained environment, just like the host based IDS.

For FPGA based systems, like an IDS, an IPS can also detect an anomaly in the system. Along with it, the IPS has additional logic based on which it can decipher the root of malfunction. For example, it tries to find out whether the vulnerability is associated with the bitstream or the FPGA fabric. If the former is the case, then it will blacklist the vendor that supplied the bitstream and send this information to all other agents to prevent any future malfunction. However, if the FPGA fabric is malfunctioning, then it will generate an alarm to replace it and, in the meantime, start outsourcing all the tasks to other agents for executing them in other resources with enhanced frequency and complete them before deadline.

5.5.3 *Analysis of IDS and an IPS with Respect to System Resources*

It may be thought what is the need to develop an IDS, when an IPS is more effective. This is done based on the criticality of the system that we are designing.

While an IDS is just associated with detection of abnormality in the system or an IDS is a detective security control, the IPS additionally tries to identify the abnormality and mitigate it if necessary and so, IPS is a preventive security control.

Thus, more logic and training are associated with the development of hardware agents that functions as an IPS than an IDS. More cost and expertise is associated with the design of IPS rather than IDS.

Thus, hardware agents working as an IPS are more costly and deployed in critical systems. While hardware agents associated with IDS are deployed in non-critical systems. More is the training and logic deployed in an IPS, better is its capability in detecting various types of anomalies and preventing them from occurring.

Hence, it is worth important to understand the system and deploy it into various criticality levels and design the IPS that can handle the various attacks related to various criticalities and deploy them.

5.6 Detection and Analysis

For generic systems, IDS and IPS modules uses the following mechanisms to detect anomaly.

5.6.1 *Signature Based*

These are essentially used for detecting the source or origin of the resource. Commonly used is the digital watermark, that is embedded in any type of image or text or video. The mechanism is pretty straight forward. If the desired watermark is not found, then it is considered that the resource arrived from another source or it was perturbed during transit. This is generally associated with pattern matching of signatures from the database.

Hardware watermarking is quite similar to digital watermarking.

Hardware IPs can be soft like simple HDL or hardware description language codes that are primarily written in VHDL or verilog, firm that are essentially register-transfer-logic (RTL) design files or fully placed netlist files or hard, which are GDSII files. These are depicted in Fig. 5.10.

Hardware watermarking uniquely identifies the IP cores in all its formats, i.e. soft, firm or hard. In soft IPs, additional secret codes are implanted, while in firm IPs, additional gates that denotes specific functionality of the source are inserted. For hard IPs, the empty spaces of the layout are filled with designer originated patterns of dummy logic that do not affect the main functionality of the design.

The mechanism of hardware watermarking involves implanting a secret functionality by the designer during its design. This is associated in the design during its development, manufacturing and fabrication. After customer buys the IP, the customer checks the originality via the secret functionality. This checking of functionality can be done via some challenge-response pairs or via some timing or power analysis or some other means.

These are non-invasive techniques of watermarking, i.e. the data cannot be altered, else the functionality of the main design will be affected. This is the mnain difference from normal digital watermarking in the image and videos, where insertion of the watermark alters the main file. Thus, the essential features of hardware watermarking can be summarized as:

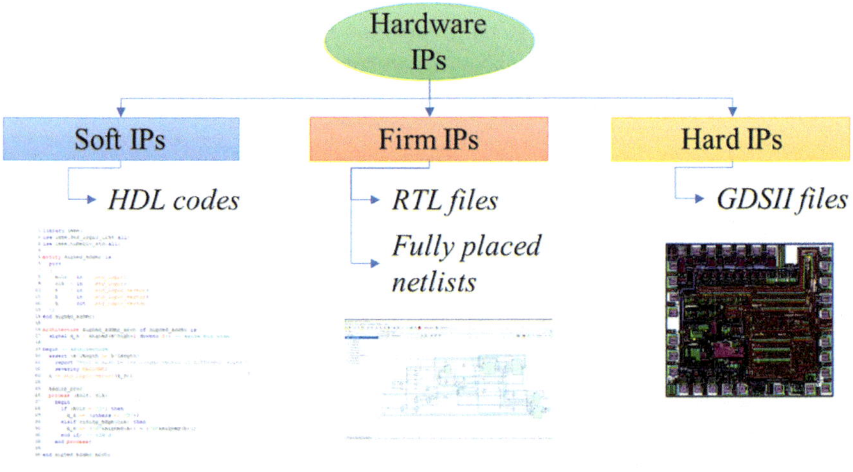

Fig. 5.10 Types of hardware IPs

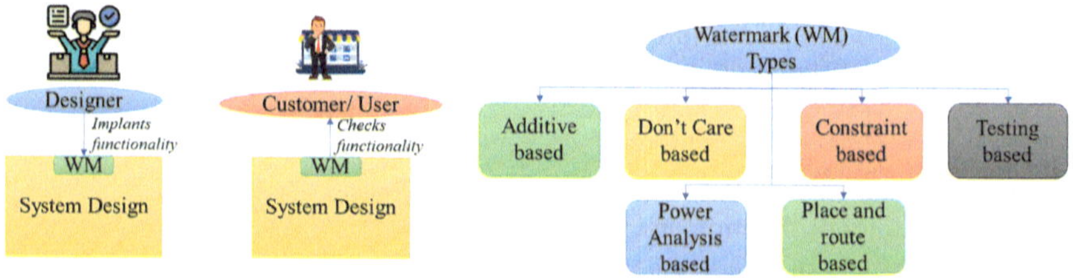

Fig. 5.11 Hardware watermarking and types

(i) must not alter the main functionality or data.
(ii) should not cause performance degradation and must be unnoticeable.
(iii) should be hard to detect and remove.
(iv) should be unique and non- reproducible.

It facilitates ease of detection and prevents piracy and counterfeiting. It also helps to trace pirated parts back to their source.

The different types of watermarking techniques are illustrated in Fig. 5.11 and also described below:

Additive Watermarking: This is done in the source code level into the functional logic and is independent from all layout constraints. Benefits include its adjustableness, difficulty to discover or remove and easy to read during cryptographic operations. However, its weakness involves high area overhead and sometimes performance degradation.

Don't Care Condition based Watermarking: This involves generating function blocks that have bogus or unneeded input combinations and outputs to these input combinations can be forced don't care conditions which can be defined during logic writing and also in finite state machines.

Constraint based Watermarking: In this, the designer's signature is a set of defined constraints and the watermarking constraints are in addition to functional constraints. More are the number of constraints, higher is the probability of authorship. Moreover, the choice of constraint must not hamper the performance. Typical examples incude unused CLB blocks and path timing constraints.

Testing based Watermarking: In this, the watermark is added to the test circuitry at the functional levels. The watermark is tested by switching to the test mode and scanning out the contents in the function mode for a number of clock cycles. These are very tough to remove. While on the flip side, it causes large area and power overhead.

Power Analysis Watermarking: In this, additional components are added to the circuit that draws power at certain frequencies. These are extracted by monitoring the power supply pin on the IC. However, dynamic power consumption magnitude will be higher at certain frequencies, which are very difficult to be understood by the adversary and very difficult to remove.

Place and Route based Watermarking: This type takes the region and grouping constraints into consideration and is mostly associated with hard IPs. This involves physical placement of standard cells and also considers row placement constraints, where some standard cells are deliberately placed in even or odd rows of the layout with a grid based abstraction. It has negligible area overhead and has a strong proof of authorship. Moreover, these type of placements cannot be altered after fabrication.

Physical Unclonable Functions: A physical unclonable function (PUF) is a hardware authentication technique, where the inherent randomness of the base semiconductor material of a device is used to generate unique responses for specific challenges, which can be used to identify the genuineness of a hardware device.

These are associated to hard IPs only and not confirm to other forms.

PUFs are used to

(i) prevent IC counterfeiting,
(ii) identification,
(iii) authentication
(iv) unique key generation.

PUF mechanism involves various stages, which are described as follows and also presented in Fig. 5.12.

Stage 1: The third party vendor implants a secret functional circuitry or physical unclonable function (PUF) in the chip during its design/fabrication.

Stage 2: Responses are recorded for different types of challenges applied on the PUF based on the physical parameters of the device like delay or power.

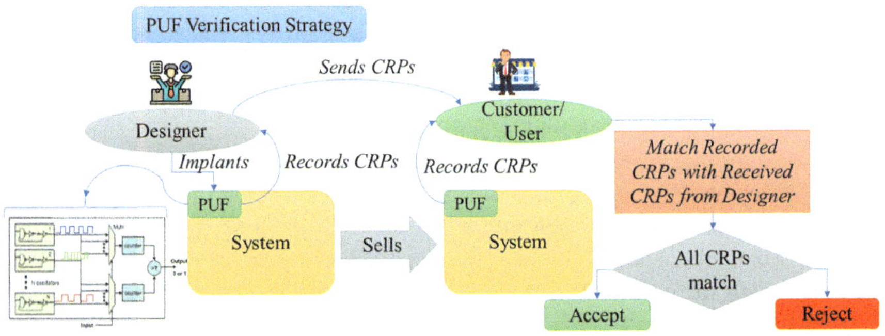

Fig. 5.12 PUF strategy

Stage 3: The third party vendor sells or provides the physical hardware device and PUF related challenge-response pairs to the user, along with the experimental parameters required for carrying out PUF based verification.

Stage 4: The user recreates the experimental environment for PUF based testing and applies the challenges on the hardware device.

Stage 5: For each challenge, the user matches the response that they obtain from the vendor and the response they obtain physically from the hardware device.

Stage 6: If both the responses are same for all the challenges, then the user can be satisfied with the genuineness of the hardware device.

5.6.2 *Anomaly Based*

This detection system is based on the type of malfunction or error that is produced. This is essentially based on the criticality of the system and what type of work is expected from the system. For instance, a IDS and IPS associated with a router is focussed on routing and network packets. While the IDS and IPS in the main system focuses on result integrity. This is mainly associated with generation of an activity profile, based on which the runtime operations are matched. This can be updated at runtime based on the evolving environment.

In general, administrators configure a device to observe or monitor its normal behavior and also to create a threshold or baseline. Then, it observes the various activities and comparison is done with the threshold or baseline. If monitored activity differs significantly to the baseline, then an alert is generated.

Anomaly-based detection systems has several names, which includes behavior-based and heuristics-based detection. However, their working is essentially the same, i.e. they try to detect abnormal activity via comparing with the baseline. The result is that an anomaly-based system often has a higher level of false positives than a signature-based system. For instance, if we are utilizing a NIDS and we need to install a software that increases network activity, then the existing NIDS might generate many false positives, if the baseline is not updated. Moreover, if the observed system is modified, the baseline or threshold must also be updated.

With the advent of AI, several agents are deployed for monitoring or tracking and updating with time. Not only such agents monitors but also predicts the various updates that will be required in future and deploy additional updates in a timely manner, as depicted in Fig. 5.13.

5.6.3 *Hybrid Mode*

A hybrid detection method is associated with combination of both signature-based and anomaly-based detection methods, as depicted in Fig. 5.14. This is quite

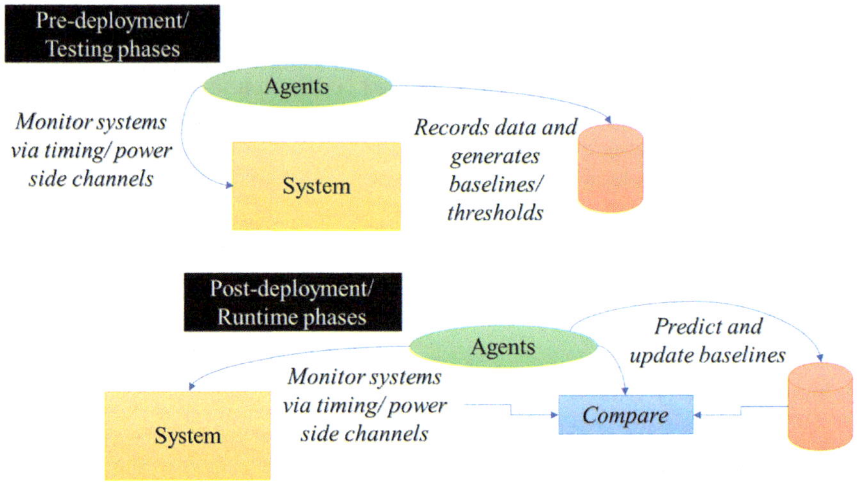

Fig. 5.13 Anomaly based detection strategy

Fig. 5.14 Hybrid detection mechanism

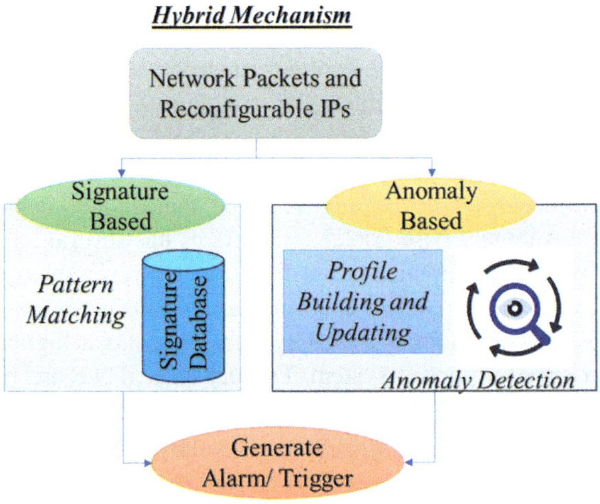

commonly used for IDSs and IPSs to increase overall performance. In this combined approach, both nature of error and the source of the resource is considered for detection.

This hybrid mode is also applicable for systems with different levels of criticality or mixed critical systems. Basic levels may require just a simple security mechanism, while the high levels may need combined security techniques.

For systems with reconfigurable hardware platforms, hardware agents that are designed to act as IDS and IPS takes the aid of various side channel parameters for detection of anomaly. For instance, the side channel parameters are time, power, etc. On board or on chip detection modules are implemented that detects the anomaly.

Moreover, the signature based mechanism is quite important as it is quite common to acquire data and resources from third party vendors. Possibilities are present where fake resources may come in and this is where signature based scheme deployment is quite important.

Data recorded may be used for training the main system and the resources may be part of some critical infrastructures. For instance, training databases may be purchased from third party vendors and several machine learning and artificial intelligence algorithms may be trained and developed based on that. If the database contains malicious data, then the intelligent algorithms developed on that, will not perform efficiently and lead to system failure.

In case of resources, malicious vendors may send counterfeit components or aged components. These will have performance degradation with time and jeopardize mission critical applications. Moreover, some components may be embedded with malicious elements like hardware trojan horses that may cause malfunction at runtime.

Another threat is modification of data or resources during transit. Man in the middle attack is imperative in any type of systems. Adversaries in the middle may intercept the data or resources and replace them with fake components and malicious data. Hence, it is important to understand whether the data and resources that are procured is obtained from the right source.

5.6.4 Whitelisting and Blacklisting

In whitelisting, the security agents try to detect abnormal activity by comparing and deciphering results with recorded values. This is done provided their current activities generates results that are below the threshold or baseline.

For instance, an application that is associated with whitelisting identifies approved data, applications or processes and for reconfigurable hardware based systems, the reconfigurable hardware intellectual properties that are associated with configuring the systems.

In blacklisting, if the results generated by an application or a reconfigurable intellectual property does not confirm to its baselines or exceeds the threshold, then it is blacklisted. Moreover, incoming applications and processes that are not in its list are also blacklisted.

For example, a switch can possess a set of system media access control (MAC) addresses that are authorized to communicate through a switch. Any systems which are not on the list or those that attempt to communicate through the path are detected and an alarm is raised for those.

5.6.5 Wireless Intrusion Detection and Prevention Systems

A wireless intrusion detection system (WIDS) works in the same manner as an IDS. Similarly, a wireless intrusion prevention system (WIPS) works like an IPS, but the application domain is different, i.e. for wireless networks. These essentially monitor or observe traffic for wireless networks and tries to detect abnormalities. These are quite effective for protecting against attacks generated from rouge access points or MAC spoofing.

Signature-based detection methods are also used over here for detecting traffic patterns from recorded types of attacks, and these also include anomaly-based detection methods. Moreover, they can even monitor key access points of the system and try to identify configuration issues.

5.7 Analyzing Results

Continuous analysis of results at runtime are the best solution. However, it will incur additional overhead in area and power. Thus, we can seek periodic analysis of the monitored data and related results. Result analysis is quite important and several tools have been developed and used to assist administrators in such an activity. These are depicted in Fig. 5.15.

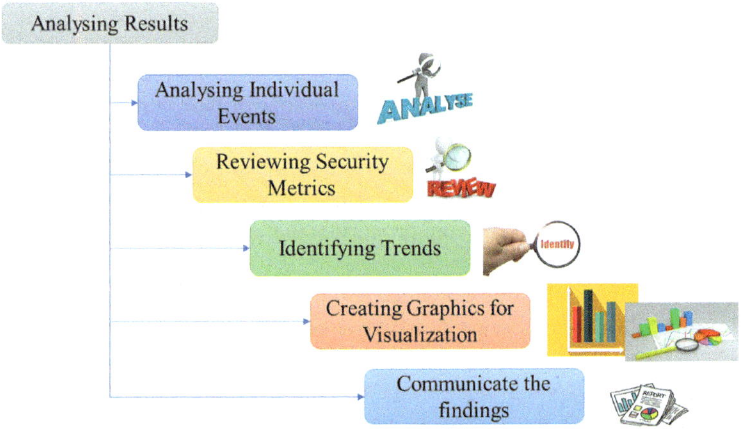

Fig. 5.15 Analyzing results

5.7.1 Analyzing Individual Events

Individual events must be analyzed to check if incidents actually occured. Administrators use different types of tools for analysis. These involve several server logs, firewalls and routers. Also log records generated by IDS or IPS are involved.

Administrators also collect several additional data like those collected by using a sniffer. This is done to capture and analyze various packets during an active attack, or memory contents of an attacked system.

5.7.2 Reviewing Security Metrics

Administrators collect suitable data from the system logs to frame information based on which suitable action can be taken.

For instance, measuring total number of false positives for a certain interval of time like a month or week is done. Such records determine whether the alert threshold is too high or too low and needs to be perfected with time.

5.7.3 Identifying Trends

Metrics help to identify a baseline or define threshold values. These can be regularly updated or changed as per the requirements.

For instance, if an organization experiences steady increase in denial of service (DoS) attacks, then analysts can compare these attacks to understand its nature and also define various performance standards related to the attacks and finally determine whether the security controls are adequate in providing security or protection.

5.7.4 Creating Graphics for Visualization

In present times, generative tools aids us in the formation of different types of graphs that helps in better understanding. Identifying trends based on graphics helps in better understanding than textual description.

5.7.5 Communicating Findings

Communicating the findings after concluding an analysis is of significant importance. Different organizations have different mechanisms to do so. Some prefer informal reports for personal use, while others can make different types of rules for its employees to adhere.

5.7.6 Detection System and Logs

Attackers or adversaries tries to avoid detection by erasing their tracks. In such a mechanism, the attackers or adversaries tries to erase log records or modify them. This is a key challenge for system designers. Investigation is necessary to determine whether the log records are modified or changed. Log records in remote systems are most valuable.

For instance, NIDS agents send copies of log records to a central NIDS management server. If attacks on network firewall or resources take place within a DMZ, then log records on the attacked resources will be compromised, but the log records in the centralized NIDS server will still be intact and used for validation and verification.

5.7.7 Detecting Unauthorized Changes

5.7.7.1 File Integrity Checkers

File integrity checkers are used by several IDS's and antivirus applications. These aid in the detection of unauthorized changes to critical system files. Such file integrity checkers uses hash techniques for detecting changes. In this, a hashing algorithm creates the hash when data is executed. The same hash will be produced irrespective of the number of times the hash algorithm works on the data. But if data is modified, then the output of the calculated hash will be different. Thus, by comparing the hash values, it can be determined whether integrity of the original life is affected or not.

A baseline is created by the file integrity checker by calculating hashes on critical system files. Recalculating the hashes and comparing these values are performed periodically. Whenever difference of hash values are detected, a change is deciphered and an alert is generated.

5.7.7.2 Unauthorized Connections

Several IDSs tries of observe various connections on systems and attempts to detect unauthorized connections. For instance, if an extranet hosts a server, which can only be accessed by a partner organization, then an IDS should detect unauthorized connections that arises from unauthorized sources.

But detection of unauthorized connections is quite difficult if the server is hosted within a DMZ and is accessible to anyone on the Internet. However, attackers generally deploy scanning tools so that it can learn information about a system. These scanning tools provide telltale signs, which the IDS can detect and send an alert.

Questions

1. Describe how an agent aids in securing a system developed with reconfigurable hardware, with the help of a diagram.
2. Describe multi-agent systems, with the help of a diagram.
3. Justify why IPS is placed before the firewall and IDS is placed after the firewall in a network.
4. State the differences between an IDS and an IPS with respect to reconfigurable hardware based critical infrastructures.
5. Differentiate between network based IPS and host based IPS and their related role for reconfigurable hardware based platforms.
6. Explain hardware watermarking and state the different types of hardware watermarking that can be done.
7. Explain the differences between digital watermarking and hardware watermarking.
8. Discuss how AI prediction can aid in anomaly based detection.
9. Explain whitelisting and blacklisting (with respect to reconfigurable hardware based systems).
10. State how result analysis is performed for security modules. Also explain the working mechanism of file integrity checkers.

References

Chakraborty N (2013) Intrusion detection system and intrusion prevention system: a comparative study. Int J Comput Bus Res (IJCBR) 4(2):1–8

Gibson D (2011) SSCP systems security certified practitioner all-in-one exam guide. McGraw-Hill Education Group

Liao HJ, Lin CHR, Lin YC, Tung KY (2013) Intrusion detection system: a comprehensive review. J Netw Comput Appl 36(1):16–24

Chapter 6
Risk, Response and Recovery

Objectives of The Chapter

- Define Risk.
- Understand threat sources, threat events, and vulnerabilities.
- Introduce the primary methods of risk management.
- Highlight risk assessment steps.
- Elaborate the incident response steps.
- Understand post incident activities.

6.1 What is Risk?

To put it simply, risk is the potential for harm to come from an attack that takes advantage of a weak spot. Things that could cause harm are called threats. When anything bad happens, it's called a loss, and a vulnerability is a weakness (Harris 2010). Risk can be calculated using the usual formula, Eq. 6.1.

$$Risk = Threat + Vulnerability \tag{6.1}$$

The connection between risks, vulnerabilities, and threats is shown in Eq. 6.1, however it is not a formal formula. Low risk is the result of few potential dangers and few weak spots. There are more serious dangers and more significant openings for attack when the risk level is higher.

A loss occurs when a threat and a vulnerability combine (a threat exploits a vulnerability). Greater dangers or vulnerabilities might easily result in increased risk-related losses. In other words, some risks have a greater impact than others.

K. Guha et al., *Cybersecurity for Reconfigurable Hardware Based Critical Infrastructures*,
https://doi.org/10.1007/978-3-031-67591-1_6

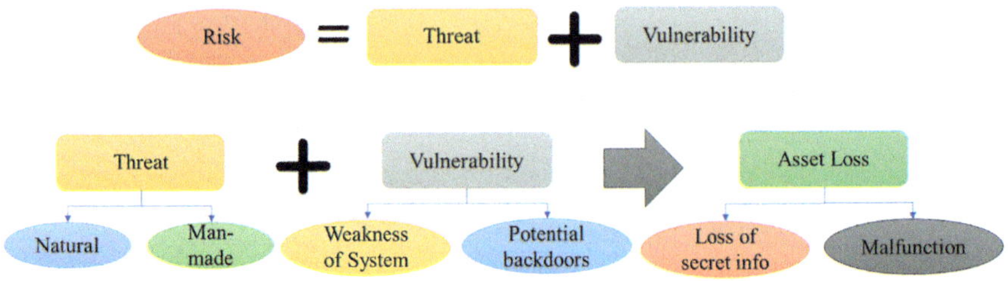

Fig. 6.1 Relation between risk, threat and vulnerability

However, reducing vulnerabilities reduces risk, resulting in smaller potential losses. Malware, for example, is a hazard to any machine. Without antivirus (AV) protection, a computer is subject to malware threats. When a virus infects an unprotected system, it has the ability to modify, destroy, or steal data, resulting in a loss for the computer's owner. The vulnerability can be mitigated or reduced by utilizing antivirus software and keeping it up to date with current antivirus signatures.

A pictorial representation of this concept is presented in Fig. 6.1.

Similarly, for reconfigurable hardware platforms, if proper authentication is not done, and the fabric is configured by a fake IP, then the system can be seized by adversaries. Moreover, bitstreams associated with hardware trojans may leak secret information or generate erroneous results or cause denial of service during a mission critical application, which may lead to system malfunction.

6.2 The Potential Threat Sources

While an attacker is commonly thought of as a threat source, there are various additional types of threat sources. They are (i) adversarial, (ii) accidental, (iii) structural and (iv) environmental.

A summary of potential threat sources is presented in Fig. 6.2.

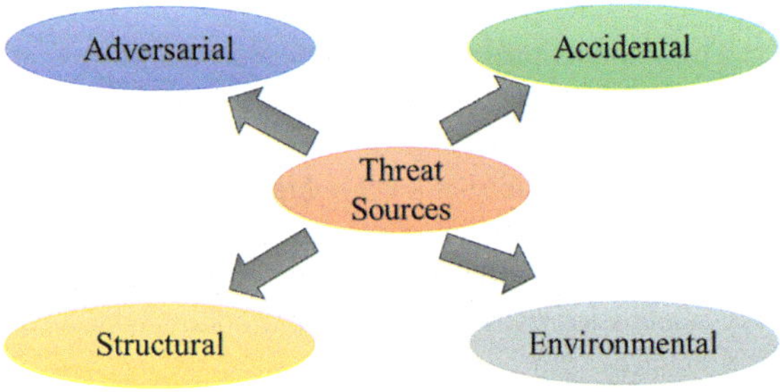

Fig. 6.2 Potential threat sources

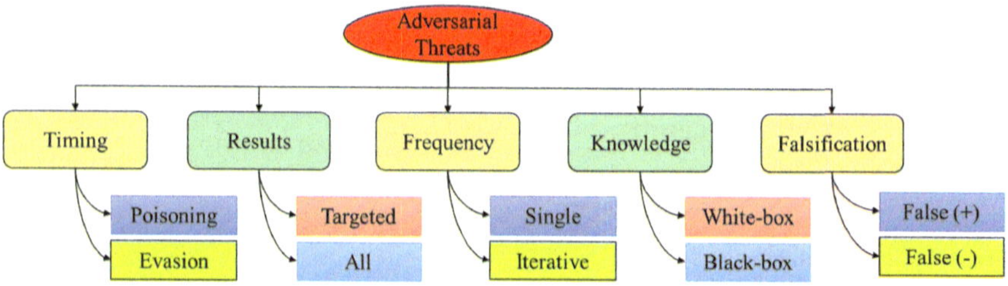

Fig. 6.3 Adversarial threat sources

6.2.1 Adversarial Threat Sources

Individuals (internal or external) or groups attempting to exploit weaknesses are examples of adversarial threat sources.

A pictorial representation of adversarial threat sources is depicted in Fig. 6.3.

Threats from within an organization come from hostile employees, contractors, or other individuals who have access to its resources and are capable of launching an attack. As far as reconfigurable hardware systems are concerned, internal threats include malicious designers making alterations to the FPGA fabric.

Threats from the outside, such as attackers in another city or country, arise from entities that do not have direct access to the organization's resources. Both loosely affiliated groups and highly organized attackers make up groups of attackers. Everyone from suppliers to competitors to partners to criminals is most often seen in groups.

Concerning reconfigurable hardware platforms, external threats can be defined as bitstreams obtained from third-party vendors that configure the FPGA fabric for different operations, along with any malfunctions related to those.

6.2.2 Accidental Threat Sources

Certain mistakes made by people during their routine tasks can pose accidental risks. For instance, a laptop can be rendered useless if someone accidentally spills coffee on the keyboard. In the event of a power outage or disruption to communication links, the availability of critical systems is compromised. Users have the potential to unintentionally cause harm to data, and the severity of the damage depends on their level of access. Regular users have limited access to resources, which helps prevent major damage, such as accidentally deleting important data. Privileged users, such as administrators, possess a higher level of access and may unintentionally cause extensive harm, such as the accidental deletion of an entire customer database. Similar to a systems analyst, it is important to acknowledge that IT security discussions often revolve around intentional threats. However, it is equally crucial to take steps to protect assets from unintentional hazards.

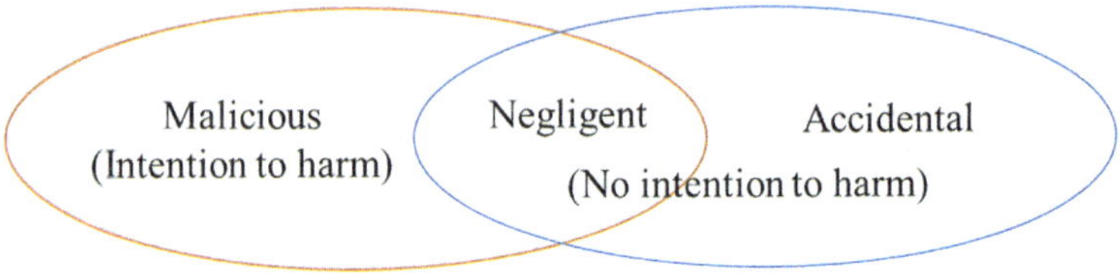

Fig. 6.4 Accidental threats

Accidental threats can either be caused by intentional harm or may have no malicious motives towards the system. In the former case, potential threats can arise from the presence of hardware trojans, which can be inserted into either the FPGA fabric or the reconfigurable intellectual properties. When it comes to the second situation, errors can occur due to various reasons, including natural causes or accidental mishandling of the system.

These are depicted in Fig. 6.4.

6.2.3 Structural Threat Sources

An example of a structural threat source might be a malfunction in software, environmental controls, or IT hardware.

For example, if a data centre's environmental controls fail, leading to a lack of air conditioning, it might cause equipment damage. Servers grow hot without air cooling, and they might overheat and fail. This leads many businesses to disable their servers in the event of an AC outage. Although the equipment is preserved, there will be an immediate decrease in availability.

For reconfigurable hardware systems, structural threats can arise due to hardware trojans implanted in the FPGA fabric during their design. Even hardware trojans in reconfigurable intellectual properties can ruin the architecture at runtime, when they are reconfiguring the main architecture. Finally, faults can occur at any time. While faults are completely unintentional, the hardware trojans either in FPGA fabric or reconfigurable intellectual properties are completely intentional malfunctions.

These are shown in Fig. 6.5.

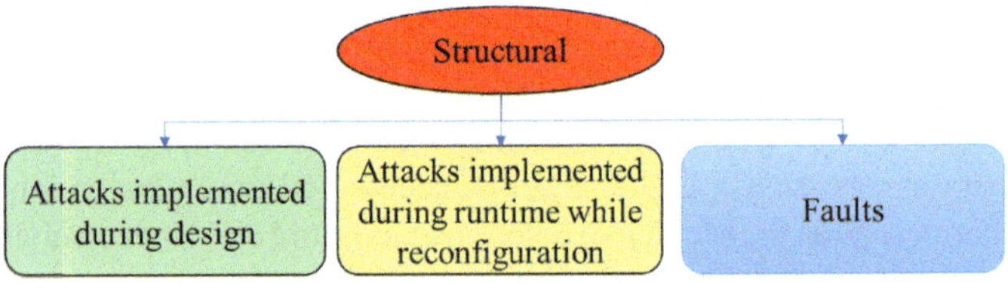

Fig. 6.5 Structural threats

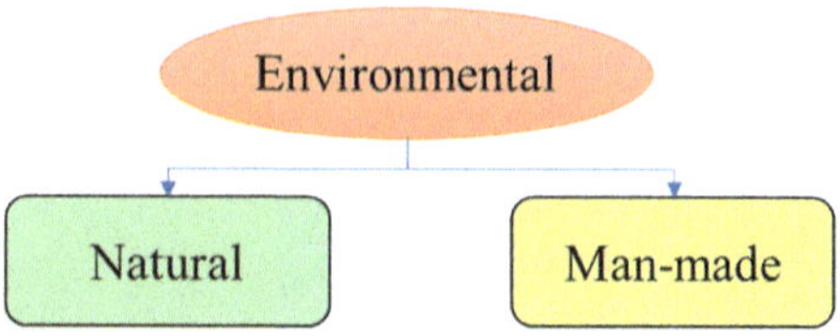

Fig. 6.6 Environmental threats

6.2.4 *Environmental Threat Sources*

Natural and man-made disasters are examples of environmental threats. Natural causes can be either meteorological (such lightning storms, hurricanes, tornadoes, and floods) or non-meteorological (like earthquakes and tsunamis). A fire is an example of a man-made tragedy. External service failures, such as a telephone link, or power outages, are also sources of environmental concern.

For remote reconfigurable hardware based systems, such environmental threats may prevent the reconfigurable intellectual properties to arrive lately that may affect the system availability.

These are depicted in Fig. 6.6.

6.3 The Threat Events

Any action or event that poses a risk to the three pillars of a secure system—its availability, integrity, and confidentiality—is considered a threat event. An example of a threat scenario would be a series of unfortunate events brought about by various threats.

Frequently, the origin and occurrence of the threat are combined into a single threat. While doing risk assessments, security professionals often disentangle the two to better identify the source of the threat and the potential action of the threat.

NIST SP 800-30 is an extensive, multi-category list of possible hostile threat occurrences. Some of the attack methods and actions that form the basis of these categories are listed below and may be seen in Fig. 6.7.

6.3.1 *Conduct Reconnaissance and Information Gathering*

In order to learn more about their target, adversaries use a variety of strategies. Information gathered using sniffers and scanners, as well as data obtained from publicly available sources, are all part of these strategies. Penetration testing is a common tactic used by attackers to infect computers with malware. If it's effective, the hacker can take control of the infected computer from afar and spy on its internal networks. These internal attacks are harder to track since most detection tools concentrate on attacks from the outside.

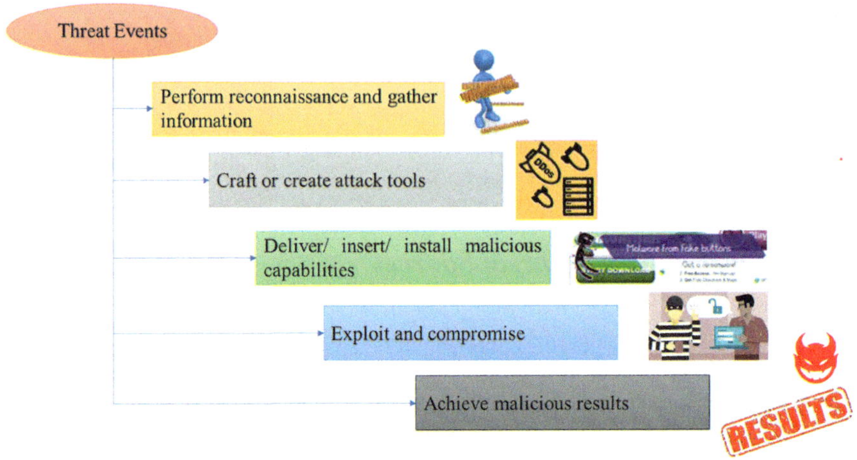

Fig. 6.7 Threat events

Similar is the case for FPGA based platforms. Adversaries from outside may send fake reconfigurable intellectual properties (RIPs), which if allowed to configure the platform, will take control of the platform.

6.3.2 Make or Craft Attack Tools

Criminals create a plethora of attack methods. For instance, they often collect information and trick users into installing malware on their systems using phishing attacks, which might include spear phishing and whaling attacks. One of their many methods has been sending bogus emails purporting to be from legitimate businesses like funeral homes and delivery services. Their goal is to attract the attention of the user. Malware is another tactic used by attackers. They often include it as an attachment.

For FPGA based systems, duplicating the origin of the RIPs is the challenge for adversaries. Adversaries tries to duplicate the watermark that is embedded in the RIPs originated from authentic sources, so that the destination cannot understand whether the RIPs is genuine or fake.

6.3.3 Deliver/Insert/Install Harmful Functionality

The attackers must discover a way to install the virus on user systems after it has been produced. It is common practice for attackers to study successful methods and develop their own versions of them. Attackers routinely include malware in ZIP files that they send as attachments with phishing emails since many consumers are ignorant of the dangers of opening them. Websites have also been their targets for malware deployment. The spyware tries to download itself automatically whenever people visit.

For RIPs associated with reconfiguring FPGA platforms, adversaries tries to understand the pattern of the secret watermarks that are implanted by the designers. For this, they may use different types of AI/ ML approaches. These AI/ ML techniques are used to understand the patterns of the watermarks and if successful, these watermarks can be embeddeed in fake RIPs that may contain malicious functionalities which will cause system malfunction or leak confidential information.

6.3.4 Take Advantage of and Compromise

Technical and nontechnical means are both exploited. To get entry to secure facilities, social engineers, for example, will tailgate closely behind permitted personnel. Methods based on technical analysis exploit vulnerabilities in unpatched systems. There are zero-day vulnerabilities that no one has ever seen a fix for, and attackers are always hunting for them. Infected systems can be remotely taken over by attackers using numerous exploits.

For reconfigurable hardware based systems, adversaries tries to take control of the system in multiple ways. Firstly, by sending fake RIPs, which when configures the FPGA platform will give control to the adversary (as discussed previously). Secondly, adversaries may try to implant hardware trojans in the FPGA fabric. Thirdly, adversaries may monitor the side channel parameters of the system to get control of the system.

6.3.5 Achieve Results (Such as Producing Negative Effects and Gathering Information)

Attackers move on to their next target after successfully exploiting systems. Inflicting system harm, entering a network to steal data, or collecting sensitive information from victims are common motivations for this.

Unfavorable implications occur whenever the outcomes compromise the security, authenticity, or accessibility of information systems and data. The attackers' next move, once inside the system or network, is to hide their tracks. The process entails evading audits and intrusion detection systems. Multiple instances of an Advanced Persistent Threat (APT) remaining active within networks for months, if not years, have been documented by the security company Mandiant.

Once a FPGA system is compromised, the users whose tasks are executed in the system may be associated with integrity attacks (erroneous result generation), confidentiality attacks (leakage of secret information) and availability attacks (generation of results after deadline, which may also be termed as denial of service). Additionally, these may jeopardize task schedules for FPGA based platforms.

6.4 The Vulnerabilities

The term "vulnerability" is used to describe any weak point in an organization, system, network, or infrastructure. A security hole exists when no control or countermeasure is in place (Gibson 2011). System and network security against Internet threats can be achieved, for instance, with the help of a firewall. An exposed network is one that does not have a firewall.

Here is how a vulnerability is characterized according to NIST SP 800-30: There is a potential for a threat to take advantage of a vulnerability if there is a hole in the information system, the security processes, the internal controls, or the implementation. The system is vulnerable if an attacker takes use of these vulnerabilities.

The following are some instances of vulnerabilities:

- Antivirus software with up-to-date virus signatures is required for computer systems. Malware can infect systems that do not have antivirus protection, resulting in severe losses.
  For reconfigurable hardware based systems, watermarks and physical unclonable functions are required to authenticate the resources. In abscence of these, adversaries may take control of the systems.
- A system can be compromised if an employee's account is not erased or disabled after termination. Either the employee or another employee could use the account to execute acts or get access to the system and cause damage. The actions of the fired worker will be mistakenly credited to the account if another worker logs in.
  Reconfigurable hardware platforms are often shared by multiple users. If a malicious user (after detection) is not blocked, then it may continue to exploit the genuine users using the same system either by causing row hammer attacks or causing denial of service.
- Making a change to a system has the unintended consequence of causing one or more difficulties. Strong change-management methods investigate prospective changes in order to predict side effects and reduce difficulties produced by changes.
  If the spaces used by a malicious user is not reset, then changes made to the spaces may cause malfunction to the genuine users who may be using the spaces later on.
- Inadequate hardening techniques Systems should be toughened or made more secure than they are by default. This involves things like changing default passwords, putting in firewalls, keeping operating systems and programs up to date with patches, and so on.
  For reconfigurable hardware systems, patches must be deployed to update the keys for the logic locking circuits or the watermarks that are deployed in RIPs. Patches for reconfigurable systems involve sending a bitstream that reconfigures the controller portion of the reconfigurable fabric.

- A complete system can go down due to the failure of a single point of failure. Disk drives, servers, communication links, power, HVAC, and even entire premises can be equipped with redundant technologies by administrators. Redundancy allows systems to handle mistakes. The disk subsystem can continue operating, for instance, in the event that a single drive in a RAID-6 array fails.
  Redundancy is a key for reconfigurable systems also. Executing a task with different bitstreams procured from different sources enhances the reliability of the system. But the key problem is associated with enhancement of overhead in area and power.
- Uninformed consumers can be duped into disclosing vital information by social engineering. Social engineers, on the other hand, are less successful when attempting to deceive informed users. Similarly, knowledgeable people are less likely to be duped into installing malware on their machines by opening attachments or clicking links in fraudulent e-mails.
  Genuine users is always under threat from adversaries who may try to steal their secret information either via side channels or via monitoring their memory access. Adversaries may even try to jeopardize the workings of the genuine users by changing the values present in the memory or may even try to cause denial of service.

Threats frequently target the system's weakest link, and once a threat exploits a weakness, the harm is done. Identifying and reducing vulnerabilities is a critical approach to minimizing losses.

The likelihood of occurrence is determined by the possibility that a threat will exploit a vulnerability. However, it is equally critical to consider the probable consequences of the risk occurring. The level of harm that could result if a threat exploits a vulnerability is referred to as the impact.

We can estimate the probable loss of confidentiality, integrity, or availability to determine the severity of the harm. In some circumstances, the impact is insignificant, hence the risk is low priority. In other circumstances, the impact is significant, making the risk a top priority.

6.5 Risk Administration

Aiming to identify, assess, and mitigate risk to a level that is acceptable to the organization is the goal of risk management. In order to effectively manage risk, it is necessary to handle both threats and vulnerabilities, which are two key components of risk. Realize that risk management is something you have to do all the time. Because threats are dynamic and ever-changing, risk management strategies need to be flexible as well.

The risk administration cycle is shown in Fig. 6.8.

It is impossible to eliminate the vast majority of risks. We can, however, lessen vulnerabilities and the impact of threats. A hurricane or an earthquake, for example,

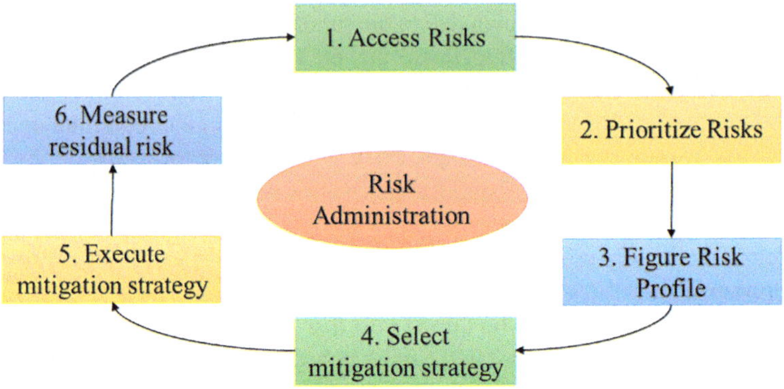

Fig. 6.8 Risk administration cycle

cannot be stopped. However, we can mitigate the impact of these natural risks by taking extra precautions, such as constructing sturdier structures or just constructing them outside of a threat zone.

Similarly, we cannot prevent an attacker from attempting to launch an attack on our system, but we may lower the adversary's chances of success by enhancing security measures and reducing vulnerabilities.

In general, a designer can choose from a number of options in response to certain threats, which can be listed as follows:

- A risk is mitigated by lowering vulnerabilities, while vulnerabilities are reduced by introducing controls or safeguards.
  For software systems, we can, for example, install antivirus software to lower the chance of malware infection. Intrusion detection systems can be added to identify and prevent network threats from entering our network.
  For hardware systems, watermarks and physical unclonable functions are commonly used to prevent risks that may be associated with using of fake resources.
- Simply refraining from doing something that puts you in harm's way is the best way to stay safe. Moving inland, for instance, can help you escape the dangers of a storm. If you would prefer to avoid the risks associated with maintaining a website on the Internet, we can opt out of hosting it.
  In general, AI and ML techniques are used to learn the different risk patterns so that they can be avoided in future. This trend is being used for both software and hardware platforms.
- It is possible to transfer or share a risk with another party. One common way to transfer risk is through insurance. For instance, in the event that a building or business is destroyed by fire, the insurance company will pay for the damage through fire insurance. Putting the onus of risk on another party is not always a workable solution. For instance, while insurance can help a business deal with some losses, it won't necessarily cover all.

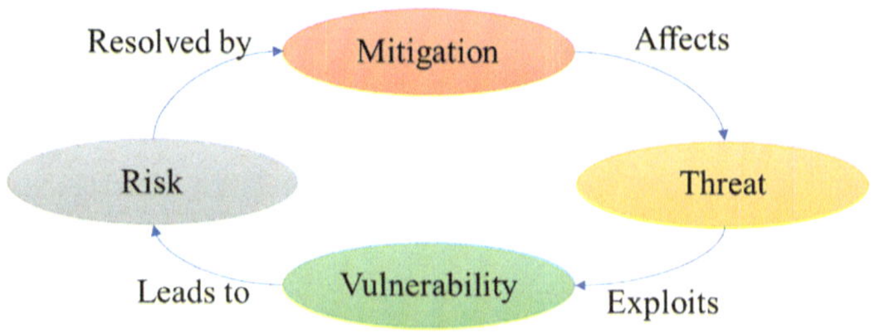

Fig. 6.9 Relationship between threat, risk, vulnerability and mitigation

For distributed systems, this is commonly followed. Agents on learning new type of threats, shares the information with other agents so that the other parts of a distributed platform may not get associated with the prevailing types of attacks in future.

- Alternatively, we can decide to bear the risk and suffer the consequences. When the value of the asset is low or when the cost of risk mitigation is more than the value of the asset, this can be justified. For instance, it's more expensive to protect mice from thievery than it is to buy the mice themselves. After taking measures to reduce the risk to a reasonable level, it is reasonable to accept the remaining risk. This is quite applicable for mixed critical systems. While the resources that are associated with some vulnerabilities may be deployed for executing low critical tasks, but it is made sure that such resources are never used for executing high critical tasks.

The relationship between threat, risk, vulnerability and mitigation is depicted in Fig. 6.9.

6.5.1 Residual Risk

A measure of the degree of danger that persists even after taking precautions is called residual risk. Put another way, residual risk is the sum of all risks minus the sum of all mitigated risks. Always check with the company to make sure the level of residual risk is manageable. In order to reduce the residual risk, the organization must take additional security measures.

An old saying states that you can never be too careful with a computer; the only way to know for sure is to never turn it on. Because, of course, the computer has no purpose here. What happens instead is that management chooses whatever controls to implement in order to bring the risk down to a manageable level. This decision-making process has reduced some risks while accepting others. Keep in mind that any residual risk is the responsibility of management in light of this. Additionally, management is responsible for losses caused by residual risk. It is up to management to choose which mitigation strategies to implement and which to scrap. Based on the needs of the system, they aim to reduce risk to an acceptable level.

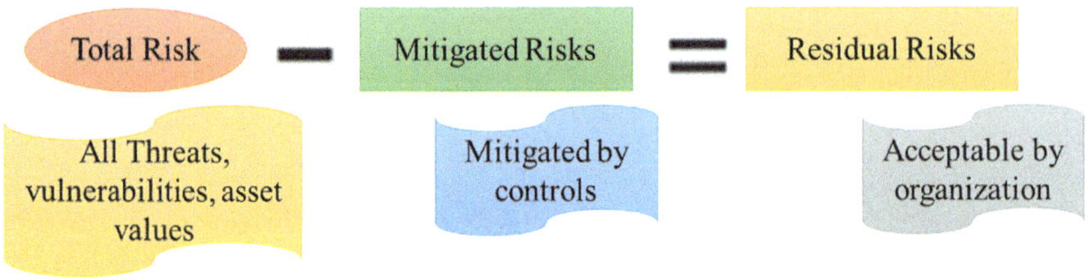

Fig. 6.10 Residual risk concept

As discussed previously, residual risk is associated for mixed critical systems. Some service providers intentionally do not replace vulnerable resources and intentionally executes low critical tasks on them, that are associated with cheap revenue. However, they deploy new resources with security features that executes high critical tasks and generates high revenue. Hence, a business strategy is in deployment.

The concept of residual risk is shown in Fig. 6.10.

6.5.2 Identifying Assets

The first and most important stage in risk management is to identify the organization's valued assets. Firms should implement controls or safeguards to mitigate risks in order to protect their high-value assets.

Despite how obvious it seems, security personnel aren't always informed of the assets the company values most. A web server, for instance, could be secretly earning thousands of dollars an hour while appearing to be no different from any other server in the server room. Server room professionals fixing a security breach can overlook the server's significance in favor of others. The valuation of assets takes into account both their physical and immaterial worth. Think about a company that makes around 6,000 *per hour selling products online. The corporation loses around* 6,000 in quantifiable revenue in the event that the website is inaccessible for one hour. Conversely, there are a lot of additional intangible losses that are harder to spot. Those customers who tried to shop on the site during its downtime may have decided to go elsewhere. These customers who have left represent intangible losses. The expense of acquiring these consumers and the potential revenue lost from not doing business with them are both included in this. It also incorporates the expense of gaining new customers. A corporation must invest in advertising to regain client trust in the event of a long-term or widely known loss. Customers who won't be impacted by the outage can opt to use a different service.

Of course, if an asset has very little value to a business, there is no reason to invest time evaluating risk. Low-value assets are generally inexpensive to replace if necessary, thus they don't require much time or effort to safeguard.

The relationship with assets is depicted pictorially in Fig. 6.11.

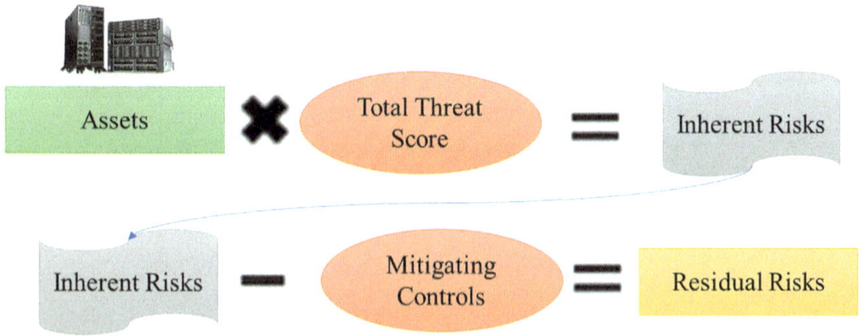

Fig. 6.11 Relationship with asset

6.5.3 *Risk Awareness and Reporting*

Risk visibility and reporting are critical components of risk management. Ideally, a business will be aware of the dangers that endanger its resources, but identifying those risks takes time and effort. Risk assessments, for example, aid in identifying these hazards and making them evident.

Antivirus, data loss prevention (DLP), firewall, and intrusion detection system (IDS) software are examples of continuous monitoring systems that keep an eye on system resources and send out alerts or reports when issues are found.

For hardware systems, low overhead self aware hardware agents are developed that observes the system's functions via side channels like timing, power, etc. and even observes the memory access rates to determine whether a system is at risk or not. If any abnormality is observed, it reports to the main controller for system debugging.

Risk awareness and reporting mechanism is shown in Fig. 6.12.

6.5.4 *Register of Risks*

To keep track of known hazards to a company and facilitate the exchange of threat information, a risk register (also called a risk log) is an essential instrument. Knowledgeable risks are documented in the risk register, which can integrate data from several sources. Examples of potential threats and the associated likelihood and impact of each are shown in an example risk register table for an e-commerce web server. In order to help security professionals prioritize risks, the risk score is calculated by adding the columns for probability and impact.

In present days, distributed ledger technology is also used (like blockchain), where the risks are recorded and updated with time.

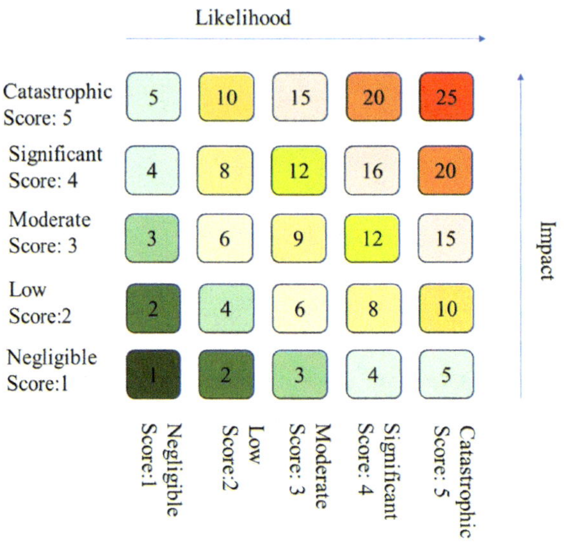

Fig. 6.12 Risk awareness and reporting

For distributed systems that employs hardware agents, the agents communicates among themselves and transfers the different types of risks that they encounter with time.

A pictorial representation of risk registration is shown in Fig. 6.13.

Attack Register	Probability (Scale 1-10)	Impact (Scale 1-10)	Risk Score (Probability * Impact)	Control
SQL Injection	1	8	8	Verify Inputs
Buffer Overflow	6	7	42	Update System
Data Breach	2	9	18	Internal Server Data Check
Side Channel	8	9	72	Synchronize Timing
…	…	…	…	…

Fig. 6.13 Risk registration

6.6 Conducting Risk Assessments

It is important to note that risk assessments are one-offs. It takes stock of the current state of affairs to identify potential dangers and potential solutions.

A risk assessment can help us identify potential threats to our company's email servers, for instance. During this assessment, various elements will be taken into consideration, such as the following: present threats, system configuration, interconnection, email processing volume, fault tolerance capabilities, power and cooling needs, and more. Any of these factors could change after the examination is over, though, and that would affect the results.

For reconfigurable hardware systems, the risk assessment involves the different RIPs that were found malicious with time, the vendors who supplied the malicious RIPs or which portions of the FPGA malfunctioned or aged with time.

Doing risk assessments has many benefits, one of which is helping security personnel to recognize and rank potential threats. As a result, the system's risk management resources will be directed towards the most pressing concerns. Either quantitative or qualitative analysis will be used in a risk assessment. Mixed critical systems involve deploying high-critical jobs to resources with no or low risk and low-critical work to resources that were previously identified as having risks.

The mechanism of conducting risk assessments is shown in Fig. 6.14.

6.6.1 *Quantitative Evaluation*

To find the real cost of a risk, a quantitative analysis uses numerical data, including financial statistics. A multitude of variables are employed in a quantitative analysis.

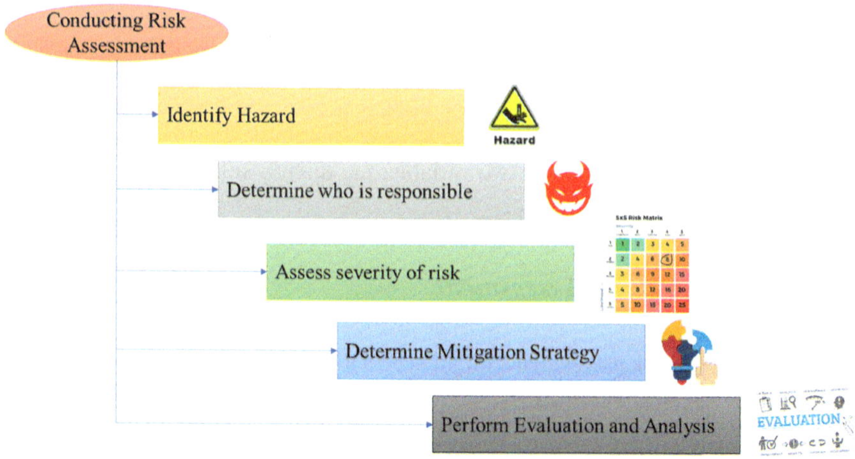

Fig. 6.14 Conducting risk assessments

Once you have the values of these variables, you can easily prioritize different risks by plugging them into a formula. Justifying the investment of controls and countermeasures can be helped by these values. Understanding these concepts is key when performing a quantitative analysis:

- "SLE" stands for "single loss expectancy", which is the actual loss caused by a single incidence. An hour of downtime for a revenue-generating website or resource could result in a $10,000 loss for the organization.
- The exposure factor is a way to quantify the potential loss relative to the value of the asset. There are instances where this is linked to SLE. Imagine, for the sake of argument, that an asset has a value of 500,000 and that insurance covers 75% of the expense to replace it. A value of 25% (or 125,000) is designated as the exposure factor.
- Based on the current environment, the annual recurrence rate (ARO) estimates how often the SLE is expected to occur in a specific year. You can figure out the ARO by looking at past data. So, assuming nothing changes, it's likely that the resource will experience three one-hour outages this year if it had two outages per year in the past.
- Multiplying SLE by ARO yields the annual loss probability (ALE). The ALE is thirty thousand dollars if the SLE is ten thousand and the ARO is three.
- The cost of control is the price that controls must be paid in order to limit or eliminate risks. A hardware RAID-6 can be used to replace, for instance, malfunctioning hard drives that caused an outage.
- If the RAID-6 is $5,000, then the control must be $5,000 as well. You have to think about the ongoing costs of these initiatives. Specialized firewalls and unified threat management systems, for instance, may necessitate further training for administrators before they can be installed and managed.

6.6.2 Analyses Qualitative

Classifying threats as low, medium, or high is a common practice in qualitative analysis, which is inherently subjective. Because the threat increases in proportion to the likelihood and impact, the two variables are directly related.

6.6.3 Steps for Risk Assessment

Projects involving risk assessments might need a lot of time and effort from the team. Contrarily, you can streamline them by sticking to a plan of predetermined stages. If different security experts are able to use the same data and come up with the same conclusions, it means the risk assessment is credible and repeatable. The ability to see patterns between tests is another benefit to the business.

The importance of reviewing previous risk assessments cannot be overstated. If they are already there on the system you are studying, you can utilize them as a starting point; in this case, we just need to find out what has changed since the last risk assessment to use them in our own method.

A service provider or a company is not compelled to take any specific steps. NIST SP 800-30, on the other hand, outlines four measures that any organization can take to achieve reliable and repeatable results.

6.6.3.1 Step 1: Get Ready for the Risk Assessment

In the first step, the company or service provider identifies resources that are used throughout the risk assessment to undertake critical tasks in preparation for the risk assessment. The purpose and extent of the evaluation, as well as its data sources, must be defined at this stage. Scope creep is a risk for any large project since it happens when the project's objectives end up being significantly broader than initially planned.

The inquiry can readily broaden into network infrastructure, including routers and firewalls, if the goal of the evaluation is to examine the hazards associated with e-mail. Providing cooling and power to the whole server room could be part of it. Like an FPGA with virtually partitioned fabric and many users given access, it might also provide malware protection for specific clients.

6.6.3.2 Step 2: Perform a Risk Assessment

During the risk assessment, we pinpoint threat vectors, threat instances, and potential weak spots. The next step is to calculate the total risk, the potential impact, and the likelihood of a risk event.

Threats from within or without might be either intentional or accidental, natural or man-made. Although it may be time-consuming and may not produce more useful data, some models suggest linking vulnerabilities with threats. It is not necessary to match threat/vulnerability pairs in many existing models (such as the NIST SP 800-30 model).

For FPGA systems, internal threats are associated with the FPGA fabric, while external threats are associated with the bitstreams procured from third party vendors. Even adversaries that indulge in side channel attacks poses external threats. External threats can also be posed by row hammer attacks caused by malicious users.

6.6.3.3 Step 3: Communicate and Disseminate Risk Assessment Findings

Those responsible for conducting risk assessments report their findings to those in charge of making decisions and other stakeholders within the company. Because they contain private information, risk assessment reports are not made available to all employees. Communication with service stakeholders is typical when results directly affect a service.

For instance, if some users are associated with a FPGA resource that needs to be replaced, then this information needs to be shared with the users and the downtime of the service is to be duly communicated to them, along with any compensations for the downtime.

6.6.3.4 Step 4: Continue to Assess Risk

You might think of a risk assessment as a brief report. To put it another way, it evaluates risk and vulnerabilities at a specific time. But, as threats are ever-evolving, it is incumbent upon businesses to endeavor to stay abreast of these changes through prudent risk management and monitoring decisions. In FPGA-based system designs, self-aware agents are entitled to continual risk assessment.

6.6.4 Address the Findings

Addressing the findings is a critical step after performing a risk assessment. In many circumstances, the assessment will highlight risks that must be managed by the implementation of controls. Naturally, the choice to address some risks is made by management. In most cases, controls are not mandated by a risk assessment. However, it is capable of analyzing and valuing controls that are advised.

Some service providers use a procedure in which they examine the outcomes of previous assessments and audit the progress. Auditing the findings afterwards provides insight into the correctness of the forecasts.

For cloud and IoT platforms that deploys self aware agents, the risk factors to be communicated to all the self aware agents. This will aid in detecting malicious vendors and possible threats and prevent future attacks.

6.7 Activity Post an Incident

Incident response makes up the final stage. The incident and the response are being looked into in this step. Determine if current plans and procedures adequately handled the issue or if response methods require improvement. It is common for service providers to find ways to enhance their current systems when they analyze their reception.

Meeting with those involved in the incident and providing questions to initiate discussions is one way to carry out activities after an incident. Having a post-incident activity meeting is necessary if the incident has a negative effect on the service

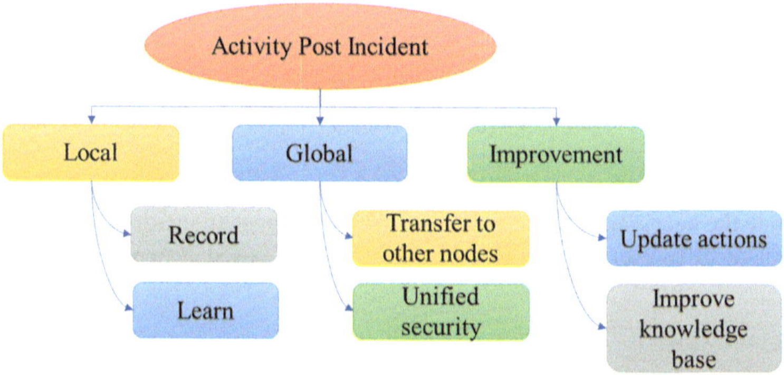

Fig. 6.15 Activities post incident

provider's mission; however, a meeting of this kind would not be fitting for a small occurrence. A post-incident report is created from the meeting's minutes and can be utilized for risk assessments and other future risk management tasks.

For reconfigurable hardware based platforms, the vendors that supplied malicious resources are blacklisted so that future malfunctions may not occur. If any deployed resources malfunctions like FPGAs or are detected with significant threats that cannot be patched, then the resources are replaced (Guha et al. 2020; Guha and Chakrabarti 2022).

Activities post incident are depicted pictorially in Fig. 6.15.

Questions

1. Explain risk with respect to reconfigurable hardware based systems.
2. Explain the relationship between risk, threat and vulnerabilities.
3. State the different types of threats and vulnerabilities and explain their relationship to reconfigurable hardware based systems.
4. Explain potential threat sources with respect to reconfigurable hardware based platforms for the following:

   (i) Adversarial threats
   (ii) Accidental threats
   (iii) Structural threats
   (iv) Environmental threats

5. Explain threat events and its related types with respect to reconfigurable hardware based platforms.
6. Explain how vulnerabilities in a system with reconfigurable hardware can be identified and related securities can be ensured.

7. Explain what is residual risk and establish its validity for reconfigurable hardware based environments.
8. State the various phases of quantitative evaluation for risk assessment and justify how they are relevant for ensuring security of reconfigurable hardware based platforms.
9. State how risks can be registered and avoided in future.
10. State the necessary activities that must be taken post an incident and justify how they are also useful for reconfigurable hardware based systems.

References

Gibson D (2011) SSCP Systems security certified practitioner all-in-one exam guide. McGraw-Hill Education Group

Guha K, Chakrabarti A (2022) Criticality based reliability from rowhammer attacks in multi-user-multi-FPGA platform. In: 2022 35th international conference on VLSI design and 2022 21st international conference on embedded systems (VLSID). IEEE, pp 234–239

Guha K, Saha D, Chakrabarti A (2020) A multi-agent co-operative model to facilitate criticality based reliability for mixed critical task execution on FPGA based cloud environment. In: 2020 33rd international conference on VLSI design and 2020 19th international conference on embedded systems (VLSID). IEEE, pp 143–148

Harris S (2010) CISSP all-in-one exam guide. McGraw-Hill, Inc

Chapter 7
Control and Countermeasures

Objectives of The Chapter

- Understanding the aspects of security controls and countermeasures.
- Deep classification of security controls.
- Inferring the benefits of cost-security analysis in the present context.
- Getting introduced to Control Goals.
- Understanding the differences between management, technical, operational, and physical controls.
- Different types of backups, such as full, differential, and incremental.

7.1 Introduction

Security controls aid in the prediction, identification, and resolution of security incidents, as depicted in Fig. 7.1. A security incident is any occurrence that compromises the privacy, safety, or availability of an entity's resources (Gibson 2011). This can happen both in the software and the hardware domains.

Loss of confidential information due to hacking or man in the middle attacks, distributed denial-of-service (DDoS) assaults, malicious software infections, power outages, and HVAC system failures are all included in this category. To lessen losses from a risk, either the defence mechanisms need to be strengthened or adversaries need to be counteracted. Administrators reduce the system's susceptibility to malware attacks by installing and maintaining up-to-date anti-virus software or deploy certain additional security monitors and attack mitigators.

Security controls can be technical and nontechnical. Technical controls integrate protective measures into systems' hardware, software, and firmware to mitigate risk

K. Guha et al., *Cybersecurity for Reconfigurable Hardware Based Critical Infrastructures*,
https://doi.org/10.1007/978-3-031-67591-1_7

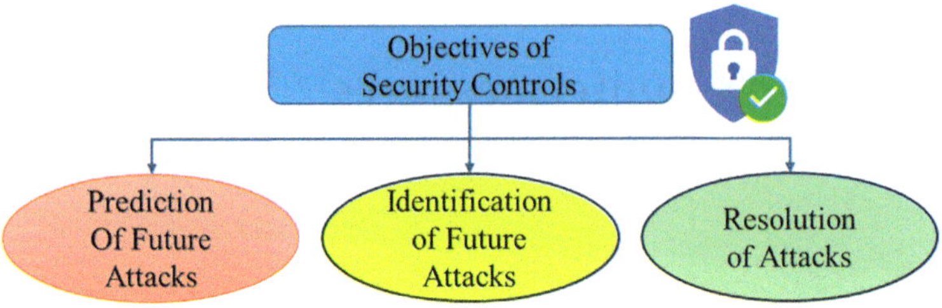

Fig. 7.1 Objectives of security controls

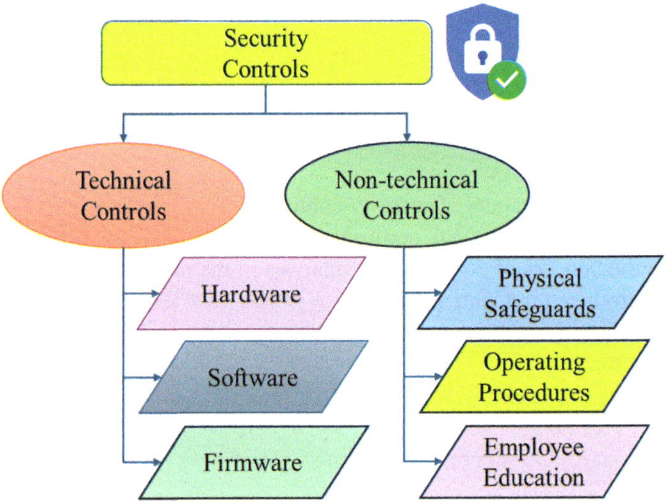

Fig. 7.2 Technical and non-technical security controls

where as Security policies, physical safeguards, standard operating procedures, and employee education are all examples of nontechnical controls. This is depicted in Fig. 7.2. Technical controls are implemented in an effort to detect and prevent assaults. An intrusion prevention system (IPS) is one such example. An example of a non-technical control is user training, which aims to dissuade users from engaging in risky behavior like opening links in unsolicited emails.

For reconfigurable hardware based systems, different types of control and countermeasures needs to be deployed. User training is of utmost importance for standalone systems, so that the reconfigurable hardware or other system components do not get jeopardized due to mishandling. While for distributed and cloud based systems that deploy reconfigurable hardware, security agents needs to be deployed that can continuously monitor runtime operations, detect anomalies or malfunction and takes necessary actions like counteracting or bypassing the threat. This is shown in Fig. 7.3.

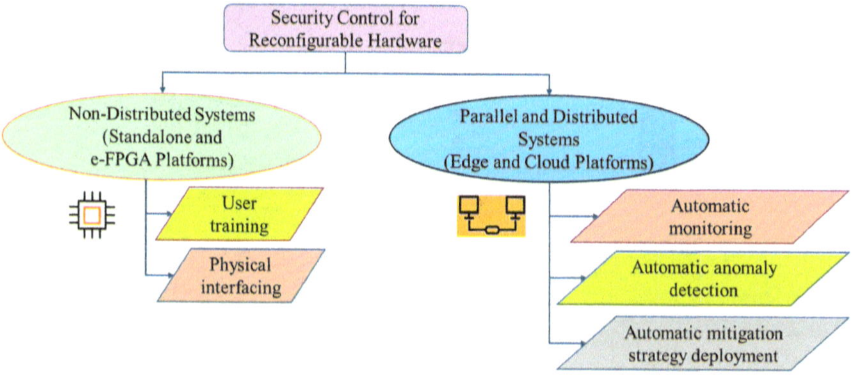

Fig. 7.3 Security control for reconfigurable hardware

7.2 Cost Benefit Analysis

Due to the finite nature of resources, organisations are compelled to evaluate and prioritise controls, investing in those that are perceived to offer the highest value (Butler 2002). In order to evaluate the value of a control to an organisation, a cost-benefit analysis is conducted, which involves comparing the costs and benefits associated with the control. In order to determine the justification for the implementation cost of a control, it is possible to conduct a cost-benefit analysis. The cost-benefit analysis additionally assesses the financial implications associated with the absence of a control measure, as well as the expenses incurred in implementing said measure.

A cost-benefit analysis is performed for every proposed control measure. Management can now make better informed decisions about the acquisition of control. The cost benefit analysis can be done using quantitative or/and qualitative research. Quantitative research utilize numerical numbers that can be subjected to a cost-benefit analysis. When making a decision regarding the implementation of a control, it is essential to evaluate both the anticipated efficacy in reducing risk and the associated costs. When determining the total cost of the control, it is crucial to include factors beyond the initial price. Before granting approval, it is necessary for the management to carefully consider and evaluate the different costs associated for implementing the controls. This is shown in Fig. 7.4.

Additional expenses include:

- *Initial Cost*: The price tag on the item or security operation. Prices may be fixed at the outset or may increase on a periodic basis (monthly or annually). For instance, anti-virus software has a one-time purchase price as well as a recurring fee for definition updates. Even security hardware has a certain lifespan, after which it ages or its functioning deteriorates.
- *Implementation Cost*: Expenses related to deployment of the control implementation expenses. Particularly expensive to implement are technical controls. Compatibility testing of the control may require personnel time and effort. It may

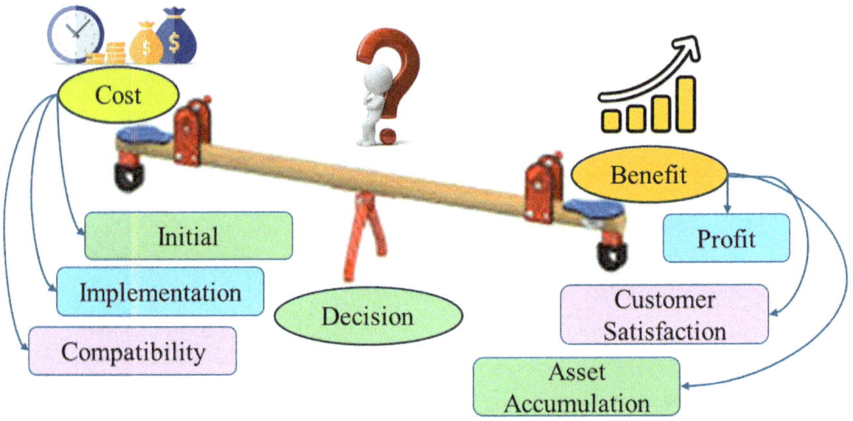

Fig. 7.4 Cost-Benefit analysis

be necessary for a vendor to do the first installation of a control, or administrators may need to undergo training prior to installing and implementing the control themselves.
For centralised reconfigurable hardware based systems, compatibility of hardware resources are needed. Such compatibility issues are quite expensive and requires significant expertise and cost.
For remote reconfigurable hardware based systems, remote configurations may be required, which are difficult to implement. Even expertise is needed. These are associated with significant implementation cost.

- *Compatibility Cost*: The expenses incurred because of the system's improved usability once a control has been implemented. Despite the fact that security should be balanced with usability, it is not uncommon for security experts to overreact and impose controls without thinking about the consequences for usability.
Both software and hardware resources and operations may not be compatible to each other. Some antivirus softwares may not run in specific hardware and vice-versa. Moreover, different security hardware may not be compatible to the main server. Checking compatibility and adhering to compatible functions and resources enhances cost than planned. Controlling productivity can have a significant impact on an organization's ability to fulfil its objective.

There is a natural progression to the lifespan of security measures. This entails doing things like picking, putting in, evaluating, and keeping an eye on security measures. When monitoring reveals a control is ineffective, security modules must continually assess the situation and choose new controls.

Selecting, implementing, analyzing, and monitoring security controls are the broad phases. These procedures are based on NIST SP 800-53, Security and Privacy Controls for Federal Information Systems and Organizations, published by the National Institute of Standards and Technology.

The various measures are:

Step 1: Decide on Protective Measures Security and then choose and assess one or more security controls for the identified information system or process. In order to determine the control's value, the evaluation often combines a risk assessment with a cost-benefit analysis.

Step 2: The next step is to put the safety measures into action. Depending on the control, this can be a simple task or a very involved one. If we want to ensure that your servers are always up and running, you'll need to do some planning ahead of time to ensure that you have everything you need to create a failover cluster.

Step 3: Thirdly, evaluate existing security measures. Personnel conducts post-implementation reviews of security measures to ensure they were properly implemented, are functioning as expected, and are adequately protecting the organization.

Step 4: Keep an eye on the security measures The efficiency of the security procedures is continuously evaluated via personnel monitoring. Changes to the protected system and external laws, rules, policies, and guidelines are among the additional elements they keep an eye on. It's important to remember that security measures have a life cycle, and that cycle never ends. The security team's job is to keep an eye out for emerging dangers and make recommendations for countermeasures in order to keep the facility safe.

7.3 Control Goals

Preventive, detective, and corrective controls and countermeasures are the most common classifications. Losses can be avoided with preventative controls, violations can be found with detective controls, and the fallout from a security breach can be mitigated with corrective controls. This is depicted in Fig. 7.5. The organization's goals may be jeopardized if sensitive information is compromised or becomes unavailable.

Fig. 7.5 Security Control Goals

Controls are implemented to avert, detect, and remediate threats to the "security triad" of prevention, detection, and correction. Some regulations satisfy all three criteria simultaneously. To stop malware from infecting a user's inbox, anti-virus software installed on their mail server might delete any attachments the sender deemed unsafe. When a user enters a malicious Universal Serial Bus (USB) flash drive into their computer, anti-virus software operating on their system will be able to detect it. Malware on a USB drive can also be eradicated with this antivirus program. Even for remote reconfigurable hardware platforms, the security modules checks the origin of the sender of the reconfigurable hardware intellectual properties before allowing them to reconfigure the platform, to prevent security breach.

7.3.1 *Preventive*

Loss avoidance is the primary goal of preventative controls. Written security policies, standards, and procedures are the most fundamental kind of preventative controls (Guha et al. 2018). Measures like division of roles, least privilege, and strong authentication are examples of preventative measures that can be found in written security policies and put into place using a variety of controls. Fences, cameras, and guards are all examples of security systems that can both prevent and discourage potential attackers. Even for hardware systems, to reduce overhead, multiple agents are deployed, which handles different types of attacks. Thus, division of roles is facilitated.

The following are some instances where preventative measures are taken (both in the hardware/software domain):

- Adhering to formal protocols by security components or agents.
- Separation of operations that are mandatory and optional to ensure obligations and the principle of least privilege.
- Robust methods of regulating access (first and foremost, robust authentication).
- Technical password restrictions that periodically require users to change their passwords and use complex ones.
- Disabling or blacklisting internal or external IP providers that were associated with malfunction.
- Different levels of security are applied depending on whether the material is considered public, private, or proprietary.
- Data encryption (both in pre and post deployment phase).

7.3.2 *Detective*

Both real-time and retrospective detective controls are available. The efficacy of preventative actions can be seen through these checks. Information can be utilized as evidence of a loss having occurred and to provide an immediate response by merging the detective control with a corrective control. The purpose of early incident detection is to allow time for a technical control or human intervention to remedy the situation. Some common types of detective controls include:

- Analyzing evidence
- Monitoring operations at runtime
- Spot abnormal functionality
- Computer virus scanners that can identify existing malware
- Records of audits (including those kept by servers and other network components, like firewalls)
- Reconciling (matching up) disparate data sources (for example, matching up a digital inventory with a physical one).

7.3.3 Corrective

A corrective control is one that responds to an event by reversing its consequences and/or re-enabling a previously deactivated or unsuccessful control. In many cases, detective controls will act in tandem with corrective controls. The corrective control is activated after an event is recognized. For instance, when antivirus signatures are updated, the program may be able to identify a live virus and isolate it. The infection will be stopped by this. A host-based intrusion detection system (HIDS) might do the same thing, discovering the host firewall is off and immediately re-enabling it. Even hardware agents tries to bypass certain threats or even mitigate them. A typical example is performing speedup of certain operations if a delay is detected, so that on the total, a real time operation never misses its deadline (Guha et al. 2017, 2019).

Some common types of regulatory safeguards include:

- Systems that detect intrusions.
- Malware detection and elimination tools.
- Data-backup and -restoration procedures.
- Contingency and recovery plans for operations.

Reconfigurable hardware platforms deploy dedicated hardware agents that perform all the three stages, i.e. preventive control, detective measure and corrective action. They are associated with observe-decide-act (ODA) paradigm. Runtime operations are observed or monitored, deciphering of operations is performed at runtime and finally, corrective actions are taken that may either be associated with bypassing or mitigation of malicious activities. This is pictured in Fig. 7.6.

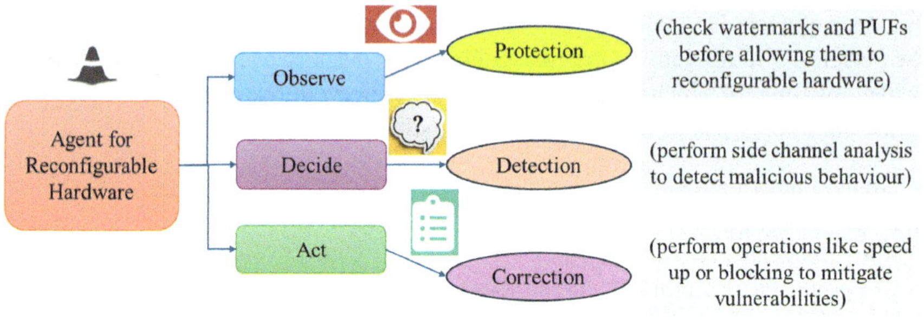

Fig. 7.6 Security control for agents of reconfigurable hardware

7.4 Other Control

We should also be familiar with compensatory, deterrent, directive, and recovery controls in addition to preventative, detective, and remedial measures. Each of the following controls might arguably fall into the preventative, detective, or corrective categories; nonetheless, these designations offer a more refined categorization:

- ***Adjustable safeguards***: These protect against a primary safeguard, when the primary one is disabled or unavailable. Let us consider a hypothetical system that implements the idea of separation of tasks, so that no single unit is responsible for overseeing the entire function. However, there will be times when one unit must take care of everything, if another one fails. A compensatory measure that can be put in place to counteract this is a series of routine reconciliation audits. The audit will look over and verify actions, and it will be able to spot any oddities. Thus, these are secondary controls, which a system uses if the primary control is impractical or undesirable.
  This is very much valid for distributed reconfigurable systems. When tasks of one of the nodes fail, then the task is taken up by another node in order to complete it. The hardware agents play a key role for communicating the failed tasks.
- ***Dissuasive Measures***: These safeguards are put in place to prevent policy violations and prevent attacks from even being attempted. A security module's presence, for instance, can discourage adversaries from sending malware. A proxy server may also show a page to users reminding them of the acceptable use policy (AUP) and disclosing that the server logs all of their online behavior when it bans access to a site from within an enterprise. This has the potential to prevent intentional page accesses by users. A deterrent control is very much like a preventative one. The key distinction is that deterrent controls aim modules to opt out of performing an undesirable behaviour. Instead of counting on components to do the correct thing, preventative controls use measures to keep them from carrying out undesirable behaviours.
  In reconfigurable hardware based systems, this is commonly done by hardware agents who checks the watermarks and PUFs of the reconfigurable IPs to determine

their sources. This prevents many malicious third party vendors from sending malicious elements.

- ***Command and control measures***: These are regulations, which are framed by authorities and must be followed. For instance, to protect patients' personal health information (PHI), businesses are required by laws like the Health Insurance Portability and Accountability Act (HIPAA). Directive controls are those put in place by an entity to ensure the safety of protected health information.
  In edge and cloud platforms, the intelligent agents takes heed of the directives and perform learning and updating as per the policies. For instance, when critical operations in hardware are performed, they are performed in separate zones and several guard bands are placed in memory to prevent attacks from adversaries, so that command and control measures can never go wrong.
- ***Recuperation Regulators***: These safeguards are designed to get things going again after an interruption or an attack that causes malfunction or failure. Reliable backups and clear instructions on how to restore them are two examples of recovery controls.
  Another typical example for reconfigurable hardware based real time systems is the speed up of operations that ensures the real time task to complete within its deadline. These are typically thought of as corrective controls, as they tries to restore the normalcy of the system.

7.4.1 Technical Security Controls

Controls that rely on hardware, software, or firmware are examples of technical (or logical) controls. The following is a list illustrating this point:

- Mechanisms for establishing a user's identify and for verifying that identity.
- Controls for auditing and accountability, which monitor actions and look for signs of suspicious behavior.
- Controls for limiting network access to system components that do not satisfy minimal security criteria.
- Controls for limiting access to systems and data to only authorized parties and blacklisting vendors who supplied components that malfunctioned.
- Confidentiality-preserving data-encryption techniques that can be effective at runtime or post deployment phases.

7.4.2 Operational Security Controls

Systems use operational controls to back up data and security policy or an organization. The following is a list illustrating this point:

- Data backup processes that consistently provide usable copies.
- Verifying the functionality of system sub-components is an example of a routine safety procedure.
- Procedures performed regularly to ensure a stable system, such as patch management software.
- Enhancing Availability through Configuration and Change Management.

7.4.3 Physical Security Control

Controls that can be handled physically are a type of security measure. They are frequently categorised with operational controls in written materials. While all security measures are crucial, physical security controls stand out as particularly vital. This category of controls includes the Physical and Environmental Protection controls listed in NIST SP 800-53. Physical access controls, as well as environmental protection controls (such temperature and humidity regulators), fall under this category.

Personnel safety should always take precedence over information technology security when planning physical security controls. Power is needed to unlock electronic door locks, for instance. Electronic door locks installed on exits should malfunction in such a manner that they unlock in the event of a power outage.

7.4.4 Integrating Class-Based Control Objectives

Control objectives (preventive, investigative, or corrective) and control classes (management, technical, operational, or physical) are frequently tested on security certifications.

The audit log is used by security components to track down security breaches. Such a control is a technological detective measure. Because it is used by security components or devices, after a security breach has already occurred and been recorded in a log, it is classified as a detective control. As it relies on machinery to keep track of data, we classify it as a "technical control".

It is not hard to link the control with the control classes (such as managerial, technical, operational, and physical) if we know the objective.

7.5 Backups

It's impossible to overestimate the value of regularly backing up our data. If a system relies on automatic controls, data will be lost in scenarios of power failure or mishandling. It is possible that the data may get corrupted due to hardware problems, software faults, inadvertent deletions, natural disasters, or even malicious attacks. A

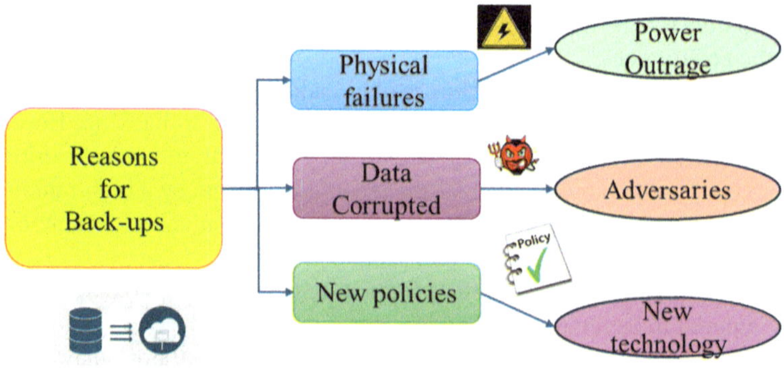

Fig. 7.7 Reasons for backups

typical example is data poisoning attacks for intelligent systems, where the learning databases are corrupted by adversaries. Even new policies may be implemented. For this, the old data needs to be stored, so that they can be used in future for various reference purposes. some of the reasons for back-ups is depicted in Fig. 7.7.

7.6 Other Control

However, having a backup system in place can make the difference between a slight disruption and a catastrophic failure. Data loss can be remedied if a backup copy of the information exists. If there is no backup, losing data will not only financial loss but also a loss of reputation for the company.

For reconfigurable hardware based systems, it is always wise to record the vendors and the intellectual properties they send for developing a system. Such system requires patching from time to time, which requires additional IPs from the vendors. If no backup of such data is kept, then in future, it won't be possible to understand whether vendor authentication was performed or not before integrating the IPs into the system.

The backup policy of an organization specifies the types of data that should be backed up on a regular basis or at periodic interval of times. The length of time an organization should preserve its backed-up data is specified in its retention policy, which is part of its backup policy. These guidelines are based on the rules and regulations framed by law and can be used to defend spending money on safety nets. The expense of tape backup hardware, media, software, and devoted backup servers is often underestimated. But if a backup policy mandates the backups, then the policy justifies the expenditures.

For instance, in reconfigurable systems, it is wise to backup the users who executed their tasks. However, the retention policy of such data may be very less than some server, as the system is resource constrained. So, the backup is only kept for a certain amount of time for resource constrained systems, as determined by rules and policies, set by the government.

In contrast, without a backup policy in place, administrators will rely on their own judgement and prior experience when deciding what data to back up. Recent data losses are often a driving factor in these choices. Administrators may prioritize user data backups based on factors such as the recent loss of data by a high-ranking user. Administrators may determine that backing up user data isn't as important if no user data is lost permanently during a year's span.

In recent times, several AI based strategies are being applied to determine which data is required to be retained and which not.

Furthermore, without a backup policy in place, administrators may not bother testing backups unless and until data is really lost. It's too late if the backups can't be relied upon. All of that information has been erased. Worse, management may opt not to sanction the acquisition of new resources when administrators seek for money to replace worn-out system resources.

A backup policy should also emphasize the value of replicating backups in a physically distinct place. In the event of a server room fire, backup resources will be destroyed along with the server. Even if backups are kept in a different part of the same building, they can still be lost if a natural disaster strikes. When backups are kept away from the original site, however, they can still be accessed in the event of a disaster.

Full backups plus differential or incremental backups are the standard for data security. While tapes are still widely used, disks are quickly replacing them as the medium of choice for these backups. Tape backups are more slower, but they save money in the long run by storing more data. Many businesses instead employ a disk-to-disk-to-tape method. This is helpful for backing up online databases, as doing so can temporarily impair database performance. Taking use of the relatively quick disk speed, administrators create backups of the data. The information on the second disk is copied to tape next.

7.6.1 Full Backups

A full backup is essentially a copy of all the data in the backup set. This works well when there is either not a lot of data to back up or when the data being backed up doesn't change frequently. You can back up the data every day if the backups are small enough.

Most backups, however, are quite sizable, meaning that both creating and restoring them takes a considerable amount of time. Strategies for full/incremental backups and full/differential backups are alternatives to full backups. Using a full/incremental backup method can help you save time when making copies of your data. You can utilise a full/differential backup technique to reduce the time it takes to recover the data. The first step in both of these plans is to create a complete backup.

7.6.2 Full/Incremental Backup Strategy

A full/incremental backup approach begins with a complete backup of the data. Then, on a regular basis, you only save new or modified information.

7.6.3 Full/Differential Backup Strategy

You can think of full/differential backups as beginning with a full backup and then periodically backing up just the data that has changed since the prior full backup.

Questions

1. Justify how the three security control objectives,i.e. prediction, identification and resolution is observed for reconfigurable hardware based systems.
2. Justify why non-technical controls are significant for standalone reconfigurable hardware based systems, while technical controls are more important form reconfigurable hardware based systems deployed in cloud and edge environments.
3. What is cost-benefit analysis? How this is valid for reconfigurable hardware based systems?
4. What are the different steps for implementing, analyzing, and monitoring security controls and how they can be related to reconfigurable hardware based systems?
5. Explain how self aware hardware agent‘s monitoring action can serve the control goal of protection.
6. State how the capabiliity of inferring of an self aware hardware agents can serve the control goal of detection.
7. Explain what type of actions can a self aware hardware agent take in order to satisfy the control goal of correction.
8. State and explain the difference between adjustable and dissuasive measures of control for ensuring security of reconfigurable hardware based systems.
9. List the various types of technical, operational and physical security controls and also state how they can be related for reconfigurable hardware based systems.
10. What are backups and state the types of backups possible. Also focus on some aspects that highlights the importance of backups.

References

Butler SA (2002) Security attribute evaluation method: a cost-benefit approach. In: Proceedings of the 24th international conference on Software engineering, pp 232–240

Gibson D (2011) SSCP Systems security certified practitioner all-in-one exam guide. McGraw-Hill Education Group

Guha K, Saha D, Chakrabarti A (2017) Self aware SOC security to counteract delay inducing hardware trojans at runtime. In: 2017 30th international conference on VLSI design and 2017 16th international conference on embedded systems (VLSID). IEEE, pp 417–422

Guha K, Saha D, Chakrabarti A (2018) Sarp: self aware runtime protection against integrity attacks of hardware trojans. In: International symposium on VLSI design and test. Springer, pp 198–209

Guha K, Saha D, Chakrabarti A (2019) Stigmergy-based security for SOC operations from runtime performance degradation of SOC components. ACM Trans Embed Comput Syst (TECS) 18(2):1–26

Chapter 8
AI/ML in Cyber-Security

Objectives of The Chapter

- Basics of Machine Learning in Cybersecurity.
- Components of ML based System.
- General Classification and Clustering Algorithms.
- AI/ML Model Implementation in Hardware Platforms.
- Objectives and Design Strategies of AI/ML Implementation on Reconfigurable Hardware Platforms.
- Attacks for AI/ML Model Implementation on Reconfigurable Hardware.
- Security Solutions for AI/ML Model Implementation on Reconfigurable Hardware.

8.1 Introduction

The application of mathematical models to data for the purpose of information extraction, pattern detection, and inference is known as machine learning (Jung 2022). There are essentially three types of machine learning models: (i) Learning with supervision (supervised learning), (ii) Learning without supervision (Unsupervised learning), and (iii) Learning with reinforcement (reinforcement learning).

Supervised machine learning employs a collection of labelled data to discern and map the concealed pattern within the data to the corresponding output label. The training label is employed to rectify or adjust the model to align with the training data as shown in Fig. 8.1. Unsupervised machine learning, in contrast, does not necessitate the use of labelled data. They acquire the skill of categorising the data points by assessing their similarity or distance to other data points, utilising the characteristics of the given data as shown in Fig. 8.2. Conversely, reinforcement learning involves

K. Guha et al., *Cybersecurity for Reconfigurable Hardware Based Critical Infrastructures*,
https://doi.org/10.1007/978-3-031-67591-1_8

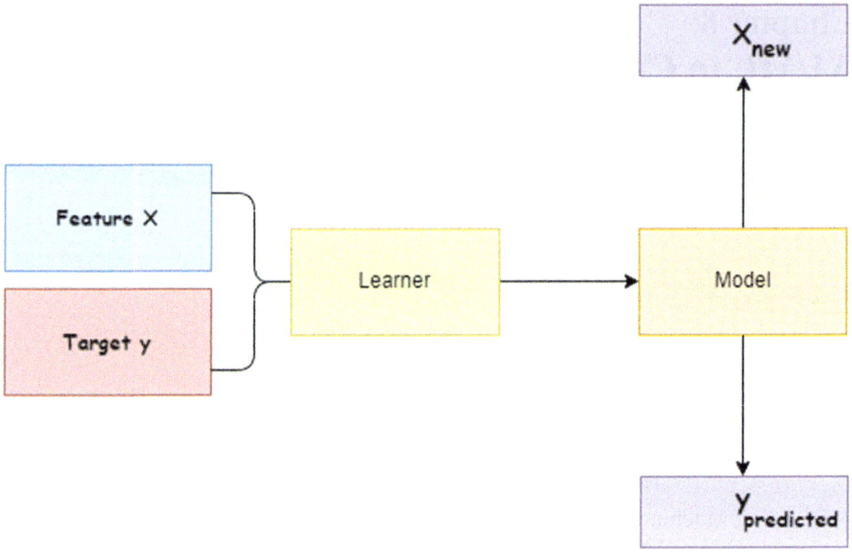

Fig. 8.1 Supervised learning paradigm

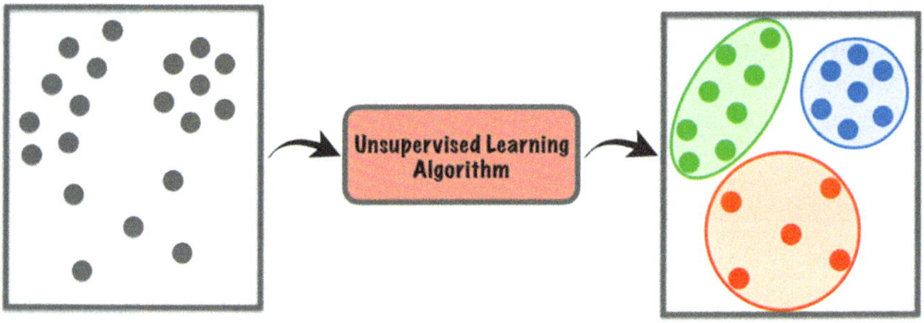

Fig. 8.2 Unsupervised learning paradigm

the use of agents that observe the environment and learn in real-time. This approach is founded on incentivizing desired conduct or imposing a penalty when undesired behaviour is observed. Therefore, a reinforcement learning agent observes and comprehends the environment, executes appropriate actions, and acquires knowledge through an iterative process of experimentation and adjustment as shown in Fig. 8.3.

Deep learning is a commonly used term that is often used synonymously with machine learning. Deep learning is a subclass of machine learning that focuses on a special type of models with multiple layers. These models learn data representations by employing simpler statistical components arranged in layers. The phrase "neural network" encompasses a broader category of layered statistical learning architectures, regardless of whether they are "deep" or not (i.e., containing multiple layers).

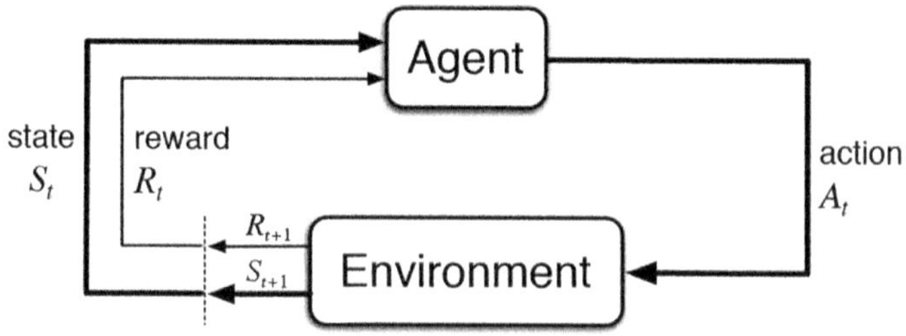

Fig. 8.3 Reinforcement learning paradigm

8.2 Machine Learning in Cyber Security

As the responsible cyber security professional, it is necessary to deploy firewalls, provide phishing training, and enforce safe coding techniques, among other actions, to prevent any security breaches (Geluvaraj et al 2019). Hence, the duty is in creating systems capable of identifying and obstructing malevolent network activity targeting any vulnerable area. In the end, these systems must determine the following:

- no files are being sent or received on the network that contain any malicious software.
- There is no breach in authentication.
- There are no instances of inbound emails containing any phishing attempts.
- No denial-of-service (DoS) attack can take place.

All of these activities are classification tasks, which require making binary determinations about the type of an observed event.

- Classify all network events as either harmful or benign.

Generic computer systems typically maintain logs containing binary files, records of login attempts, incoming emails, and outbound requests. On some occasions, we may have knowledge of previous attacks and be able to establish a connection between them and the associated log entries. To initiate the process of fixing the issue, we conduct an analysis of historical data in order to identify any discernible patterns that may suggest the presence of malicious assaults. If a single IP address sends more than 20 requests per second to your servers within a 5-minute timeframe, it is highly likely that a Denial of Service (DoS) attack is occurring. Once patterns in the data have been identified, the subsequent task is to convert these patterns into an algorithm. An algorithm is essentially a function that takes input data related to the subject being classified and produces a binary response: either "malicious" or "legitimate".

One more area that machine learning can help with is regression problems. These involve trying to predict the value of a real-number variable. For instance, considering factors like job role, access level, length of employment, security practices, etc., we would like to predict the number of phishing emails that each employee receives in a given month. Time series analysis is a popular approach to regression problems that include inputs with a temporal dimension. Predicting the amount of account sign-in's from the Seattle office from its past records is one example, as is estimating the future value of a stock based on its historical performance. An additional part of regression analysis is anomaly detection. Finding cases where a projected value differs greatly from an observed one, indicating the presence of an anomaly, is what this process is all about.

Furthermore, machine learning is employed to address clustering challenges, which involve determining the similarity between a set of data points. For example, when analyzing a substantial dataset of website traffic, it may be desirable to determine the clustering of requests. Clusters may correspond to botnets, mobile service providers, and legal consumers.

Supervised machine learning makes use of labels on historical data and attempts to predict labels on future data. You can train a spam classifier to predict if an incoming message is spam by applying it to a large dataset of emails labeled as spam or ham. In contrast, unsupervised machine learning can be employed when labels are absent from historical data. Predicting unknown labels is the goal here, like differentiating between a network attack by an unknown number of botnets. Tasks involving clustering are part of unsupervised learning, whereas supervised learning includes classification and regression.

8.3 Components of Machine Learning Based System

A machine learning algorithm's basic function is to build a model from a training dataset. The model is a computational framework that generates predictions by feeding in new data points that match the structure of the training data. A definition of a machine learning algorithm always includes these three interrelated parts. You can think of all the available models as belonging to a single model family.

- A loss function, which permits quantitative comparison of various models
- An optimization technique that enables us to choose the optimal model within a specific family

The block diagram of machine learning based systems is depicted in Fig. 8.4.

8.3.1 Model Families

Each instance of fraud is represented as a feature vector in a fraud detection instance that uses features extracted from the dataset. Decision boundaries, which can be generated by ML models, are surfaces in vector space that differentiate between

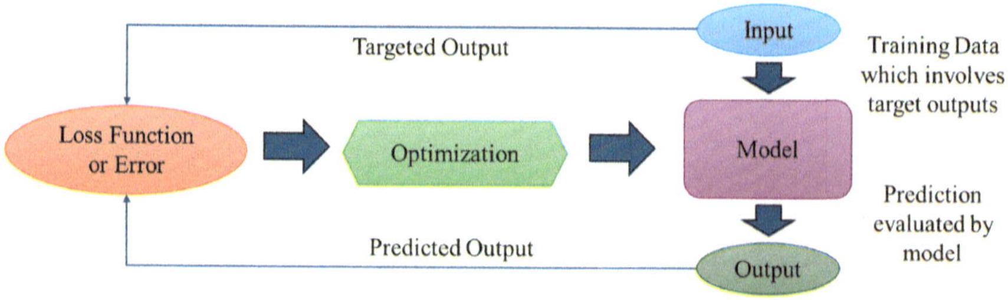

Fig. 8.4 Block diagram of machine learning based systems

fraudulent and legitimate examples. The points are classified as either valid or fraudulent by the decision border. Boundaries can be as simple as lines or, in higher dimensions, hyperplanes, or as complex as a network of coupled nonlinear areas.

Instead of assigning each point in the vector space as either "fraud" or "nonfraud," we have the option to assign each point a likelihood of being fraudulent, allowing for a more detailed analysis. Our machine learning method produces a function that assigns a value between 0 and 1 to each point in the vector space. This value represents the chance of fraud in our specific case.

Each machine learning method has a constraint in its capacity to identify a particular decision boundary or probability function, which can be defined by a finite set of model parameters. A hyperplane in the vector space is the most basic form of a decision boundary, representing a linear separation between different classes.

8.3.2 Loss Function

After we have narrowed down our selection of prediction algorithms to a specific parametrized family, it becomes crucial to identify the most optimal one for the given training data. An optimal algorithm is one that effectively maximizes or minimizes a particular value obtained from input data. This is typically referred to as an objective function. The objective function in machine learning is also known as a cost function or loss function. It measures the "cost" or "loss" associated with inaccurate predictions. A loss function is a mathematical function that assigns a numerical value to the predicted labels and the actual labels. Using a machine learning method, we strive to discover the best model parameters that produce projected labels for the training set while minimizing the loss function. In regression scenarios, the typical loss function used when the prediction algorithm produces a continuous numerical value instead of a categorical label is the sum of squared differences. This loss function is applicable to classification problems as well. In such cases, the algorithm's output probability estimate, denoted as y_i, can take values of either 0 or 1. This is depicted in Fig. 8.5.

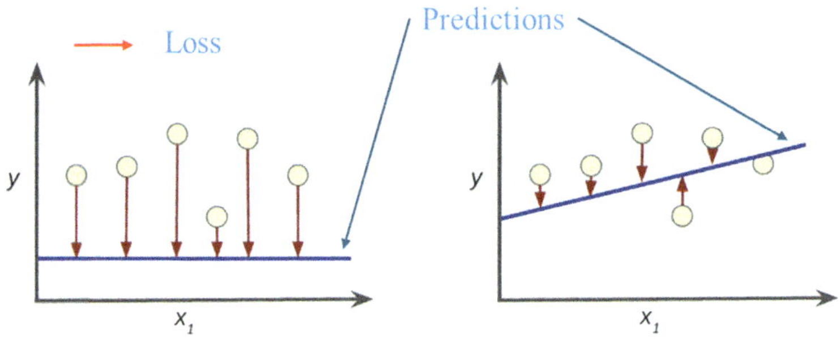

Fig. 8.5 Depiction of loss functions

8.3.3 Optimization

In the final stage of machine learning, the main goal is to find the optimal set of parameters that minimizes the loss function. A search is conducted using an optimization algorithm. Throughout the process of fitting your machine learning model, you have the chance to explore a wide variety of optimization methods. The primary goal of an optimization algorithm is to minimize or maximize an objective function. In the world of machine learning, the objective function is carefully designed by taking into account the model's parameters that can be adjusted during the learning process. The main goal is to find the values of these parameters that maximize the objective function.

8.4 Popular Classification Algorithms

ML is associated with several classification algorithms (Gama and Brazdil 1995). Having gained a comprehension of the fundamental workings of machine learning algorithms, we will now provide a concise overview of the prevailing supervised learning classification techniques.

8.4.1 Naive Bayes

The Naive Bayes classifier is considered one of the earliest statistical classifiers. The classifier is referred to as "naive" due to its reliance on a statistical assumption that features are randomly selected from an undisclosed distribution. It's quite uncommon for this assumption to hold true in real-world situations. Picture a spam classifier that depends on the words used in a message as its features. According to the Naive Bayes hypothesis, a spam message is generated by randomly selecting words, where each word has a probability p_w of being chosen as spam. The same principle applies to positive messages. This assumption is clearly ridiculous; to begin with, it completely ignores the order of words. Although the assumption is incorrect, it has been proven that Naive Bayes classifiers are extremely efficient for tasks like spam classification.

Naive Bayes is based on the following basic principle: With a data point and a set of features $X = x_1, ..., x_n$, we want to determine the likelihood that the label Y belongs to the class C. A conditional probability $P(Y = C/X = x_1, ..., x_n)$ represents this idea.

Using Bayes' theorem, this probability can now be expressed as

$$P(Y = C/X = x_1, ..., x_n) = \frac{P(X = x_1, ..., x_n/Y = C).P(Y = C)}{P(X = x_1, ..., x_n)} \tag{8.1}$$

Assuming that samples within each class are selected independently, we can reduce the probability of $X = (x_1, ..., x_n)$ to $P(X = x_1)$. Determine the likelihood of X having values between x_2 and x_n. Figure 8.6 depicts the likelihood, class prior probability, predictor prior probabilty for a posterior probability for Naive Bayes learning.

By utilising labelled data, we can calculate the probability of $Y = C$, which represents the ratio of samples in class C to all labeled samples. For two-class classification, we can rely on the ratio of the probability estimates for the two classes $C1$ and $C2$ without having to calculate the denominator.

A diagrammatic representation of the working of Naive Bayes Classifier is depicted in Fig. 8.7.

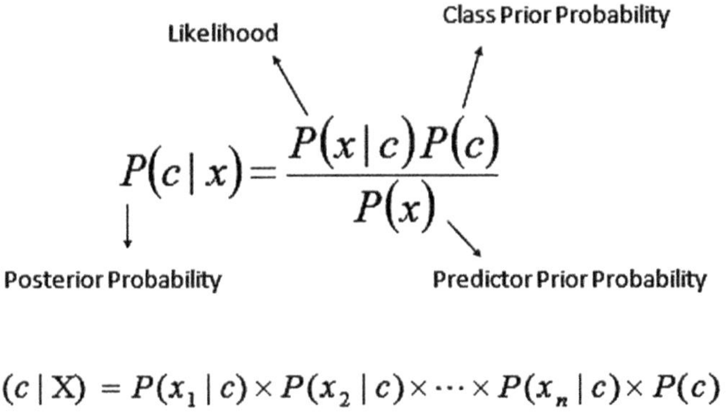

Fig. 8.6 Naive Bayes learning with different probability calculation formula

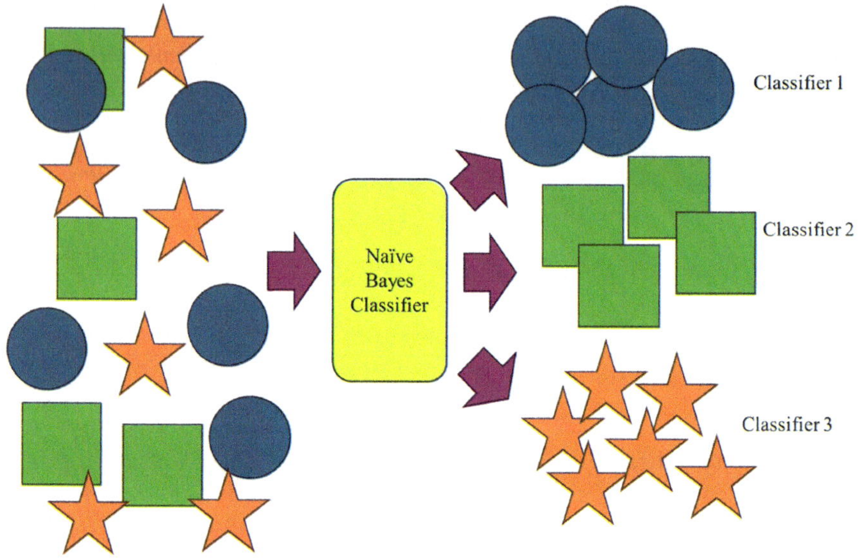

Fig. 8.7 Diagrammatic representation of working of Naive Bayes classifier

8.4.2 Logistic Regression

Logistic Regression is a statistical model that takes numerical feature vectors as input and aims to forecast the logarithm of the odds of each data point occurring. These logarithmic odds can then be converted into probabilities using the sigmoid function. As the decision boundary in the log odds space is linear, the model's output score will either increase or decrease monotonically when the value of a feature is increased, depending on the sign of the coefficient. Logistic regression is a highly popular algorithm in practical applications because of its various advantageous properties. It can be trained efficiently and distributedly, making it suitable for large datasets with millions of features. Additionally, it enables a concise description and fast scoring algorithm, using a simple dot product. Moreover, logistic regression is explicable, as it allows for the computation of the contribution of each feature to the final score.

A diagrammatic representation of the working of Logistic Regression Model is depicted in Fig. 8.8.

8.4.3 Decision Trees

One of the most important qualities of versatile supervised learning models, decision trees are also very easy to understand and work with. A decision tree is a binary tree data structure that is utilized for making conclusions. Trees are frequently used for

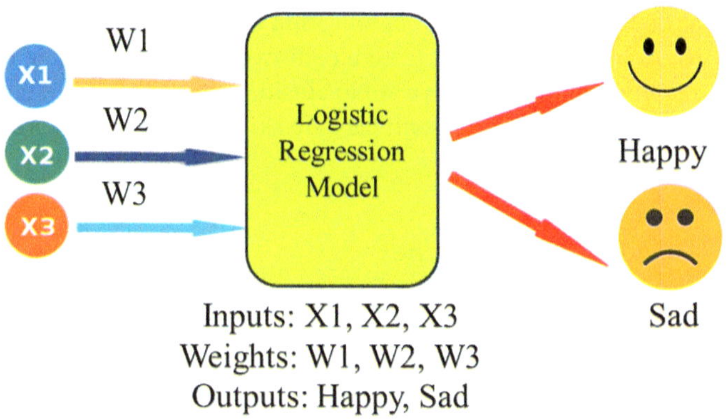

Fig. 8.8 Diagrammatic representation of working of logistic regression model

visualizing and interpreting data in several fields, not limited to machine learning, because to their intuitive characteristics. Their popularity in machine learning stems from their ability to predict both categorical and real values, known as classification and regression trees, respectively. Additionally, they can handle numerical and categorical data without requiring normalization or the creation of dummy variables.

Regression trees are commonly employed when the dependent variable is continuous. Variance reduction is the overall decrease in variance that occurs when a set is divided into two subgroups. The split that achieves the greatest reduction in variance at a node in a decision tree is referred to as the optimal split. Figure 8.9 depicts a decision tree for classifying weather conditions.

The method results in a hierarchical structure where each node represents a binary choice, its descendants represent the two possible outcomes of that choice, and each terminal node represents the class of data points that follow the path from the root

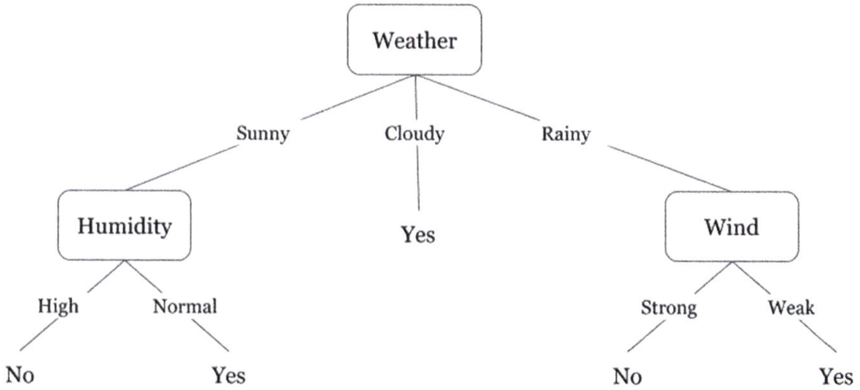

Fig. 8.9 An example decision tree

to a terminal node. Every result of classification or regression can be represented as a set of Boolean conditions that trace a path from the starting point to the final nodes of a tree. It is quite straightforward to elucidate the outcomes of decision tree classification or regression. As a result, they are skilled at managing large datasets.

8.4.4 Support Vector Machines

An SVM is a highly effective tool utilized in machine learning to classify data by constructing a hyperplane in vector space. The hyperplane efficiently separates the two classes identified in the dataset. Understanding the loss function is a key differentiator between logistic regression and SVMs. Logistic regression utilizes a log-likelihood function that takes into account the accuracy of the estimated probability, penalizing all data points, including those that are correctly classified, based on their inaccuracy. On the other hand, a Support Vector Machine (SVM) utilizes a hinge loss function that specifically penalizes data points that are either misclassified or located very close to the decision boundary. The SVM classifier aims to find the hyperplane that maximizes the margin between the two classes. Understanding the margin is crucial as it signifies the distance between the separating plane and the nearest data points on both sides. When the data is not easily separable, points that fall within the margin are penalized based on their distance from the margin.

The kernel trick is a powerful mathematical technique that transforms a simple linear decision boundary into a more complex and flexible nonlinear one. This clever method greatly enhances the capabilities of Support Vector Machines (SVMs). At its core, a kernel transforms one vector space, V_1, into another vector space, V_2. The kernel is a mathematical function that is defined on V_1 and V_2, and it is denoted as $K(x, y)$. Every element x in V_1 is associated with the function $K(x, y)$. The vector space V_2 consists of all these functions.

The diagrammatic representation of working of support vector machines is shown in Fig. 8.10.

8.4.5 K-Nearest Neighbour

One of the best examples of a lazy learning algorithm is the k-NN algorithm. This ML method does not do most of the calculations during training but rather waits until classification. Lazy learning models do not develop broad understandings of data during the training phase. On the other hand, they gather and save all the training data points given to them, utilizing this data to generate precise conclusions about the test sample while classifying. The k-NN method is widely recognized as a fundamental machine learning algorithm.

Throughout the training phase, the model retains all feature vectors and sample labels. In order to make a classification prediction, we need to look at the labels of the

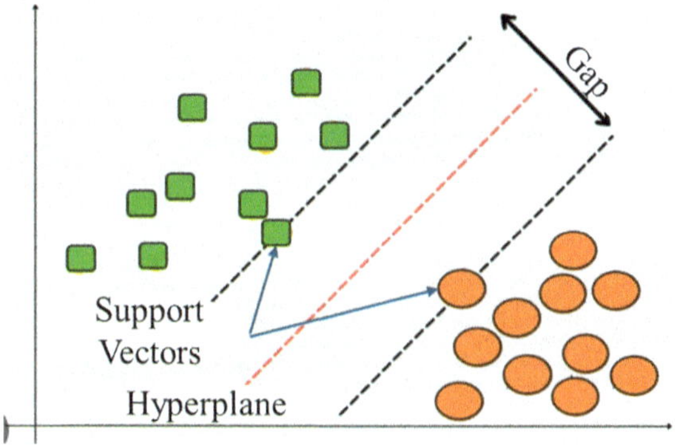

Fig. 8.10 Diagrammatic representation of working of support vector machines

k nearest neighbors of the test sample and select the one that appears most frequently. Finding the proximity of two points requires the use of distance calculations. If the variables are continuous, the distance can be the Euclidean one; if they are discrete, the distance can be the Hamming one.

The k-NN algorithm's training phase is usually faster than other learning algorithms. Nevertheless, this rapidity comes with the trade-off of longer processing time for classification.

Although K-NN is often used as a simple example to introduce beginners to the concept of machine learning, it has limited practical applications in real-world scenarios.

- Models must have large sizes to accommodate all the feature vectors and labels from the training data, at the very least.
- The classification process is hindered by the delay in generalization work until the moment of classification. Locating the nearest neighbors can be a time-consuming task, especially when the model stores training data points in an inefficient spatial search format.
- Efficiently adapts to class imbalance in the dataset. The classification results may be influenced by the imbalance in the number of samples across different classes in the training data. This is because samples from these categories have a higher probability of being included in the k-NN set of any given test sample.
- The accuracy of the classification has been affected by the presence of features that are noisy, redundant, or unscaled.First item on the list. Determining the value of the parameter k can be quite a challenge. Choosing the right value of k for classification results can be quite challenging, as it needs to perform well across all regions of the feature space, which can have diverse densities in the dataset.
- It struggles to function efficiently in high dimensions because of the "curse of dimensionality". In addition, as the number of dimensions in the feature space increases, the "neighbourhood" around a specific point expands, resulting in a less predictable selection of neighbors.

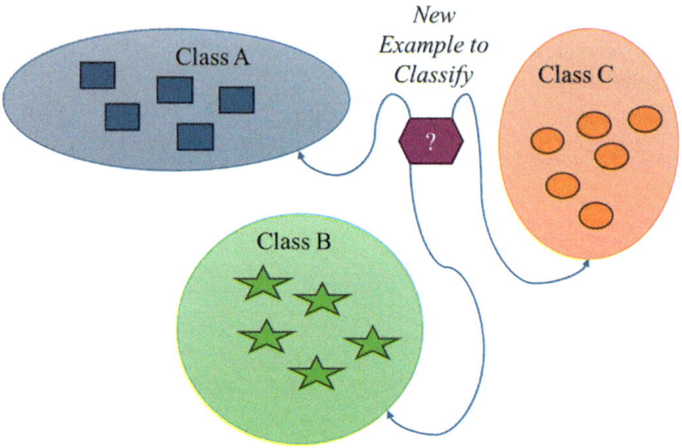

Fig. 8.11 Diagrammatic representation of working of K-nearest neighbours

The diagrammatic representation of working of K-Nearest Neighbours is shown in Fig. 8.11.

8.4.6 Artificial Neural Networks

Artificial neural networks (ANNs) have become increasingly popular as a machine learning technique. In 1942, McCulloch and Pitts published a groundbreaking paper that delved into the inner workings of neurons in the human nervous system, laying the foundation for the development of neural networks. The rise in popularity of neural networks can be attributed to the advancements in hardware, particularly graphics processing units (GPUs), which have greatly accelerated and parallelized the training of artificial neural networks (ANNs). Additionally, the abundance of data available has played a crucial role in enabling ANNs to excel in complex tasks like image and speech recognition. The human brain consists of an immense number of neurons, approximately 10 billion, with each neuron connected to tens of thousands of other neurons. Neurons receive electrochemical inputs from other neurons, and when the sum of these electrical inputs surpasses a specific threshold, the neuron transmits an electrochemical signal to its connected neurons. If the input is below this threshold, the neuron remains inactive. Every neuron functions as a basic processing unit with minimal capabilities. However, when these neurons are interconnected in different patterns and layers, the brain can accomplish intricate tasks, such as differentiating between a cat and a dog or comprehending profound philosophical ideas.

Artificial neural networks (ANNs) originated from endeavors to replicate the functioning of neurons in the brain, with the aim of attaining learning capabilities akin to those of humans. Neurons were represented by mathematical step functions,

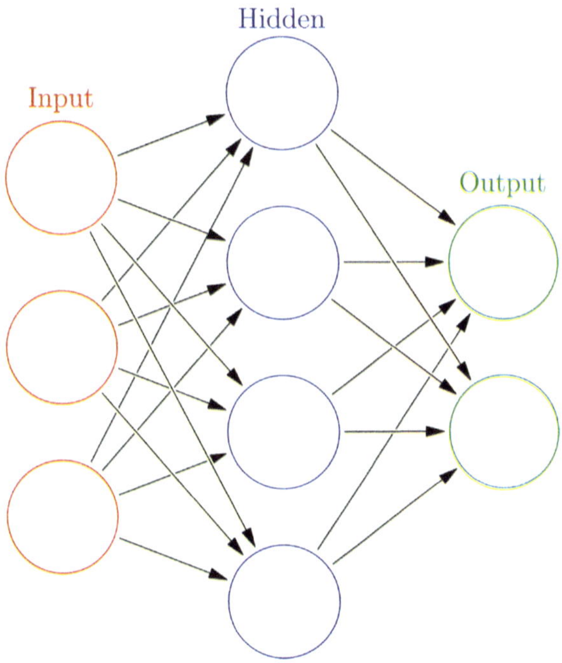

Fig. 8.12 An example neural network

known as activation functions. These functions received input from certain neurons, and when activated, emitted output to other neurons. This mathematical model of a biological neuron is commonly referred to as a perceptron. By manipulating the model's topology, activation functions, learning objectives, and training methods, we can generate a wide range of neural networks using perceptrons. Layers of interconnected neurons constitute the bulk of neural networks. Figure 8.12 depicts a generic structure of a neural network. Neurons in one layer take data from neurons in the layer below them and, when activated, send signals up to the layer above them. Weights are assigned to the connections between neurons, while biases can be assigned to individual neurons or layers. Here are the parameters that will be trained through the backpropagation process, which we will provide a brief description of. Prior to the start of the simulation, all the weights and biases are initialized randomly. We perform two steps for every sample in the training set:

- Perform a forward pass. Input the data into the artificial neural network to get the current prediction.
- Backward traversal. When the prediction turns out to be accurate, it is important to acknowledge and strengthen the connections that contributed to it by adjusting their weights based on the level of confidence in the prediction.
- If the prediction is wrong, penalize the connections that played a role in the incorrect result.

Neural networks have been extensively utilized in various industries and are backed by a substantial amount of academic research. There is a wide variety of neural network infrastructures available for both supervised and unsupervised learning. Certain artificial neural networks possess the capability to autonomously learn features without the need for extensive manual engineering.

8.5 Clustering Algorithms

Occurrences of negative events often tend to happen in groups. When someone tries to breach your network, they often make multiple attempts before they can succeed. On the other hand, if someone wants to deceive a significant number of people with pharmaceutical spam, they will have to send a considerable volume of emails. By categorizing your traffic and effectively blocking malicious actors, you can streamline your defense efforts and make your job as a defender more manageable. This process of segmentation is referred to as clustering.

Now, let's delve into a few widely used data clustering techniques.

Clustering's geometric intuition is quite simple: the goal is to group data points that are in close proximity to each other. Therefore, in order for an algorithm to be effective, it is necessary to have a specific method of quantifying "closeness", which is commonly referred to as a metric. The choice of metric and clustering algorithm depends on the nature of your data, whether it is in the form of real-valued vectors, lists of items, or bit sequences. Several clustering algorithms are present (Xu and Wunsch 2005). Let's now explore the algorithms that are widely used.

8.5.1 k-means Clustering

When it comes to clustering, the k-means algorithm is often the initial choice. When you have a specific number of clusters in mind, the k-means algorithm can be applied to real-valued vectors. The variable k represents the number of clusters. Our main goal is to ensure that every data point is assigned to the most appropriate cluster, minimizing the overall distance between each point and its cluster centroid. For this purpose, the distance is calculated using the standard Euclidean distance in a vector space.

$$d(x, y) = \sqrt{\Sigma_i (x_i - y_i)^2} \tag{8.2}$$

The k-means algorithm calculates a cluster assignment function f: X → 1, ...*k* that minimizes the loss function in mathematical terms:

$$L(X) = \Sigma_i d\left(x_i, c_{f(x_i)}\right) \tag{8.3}$$

where X is the dataset, c_j is the jth centroid, and d is the distance between two points. The value $L(X)$ is referred to as "inertia". The following is the standard algorithm for computing k-means clusters:

- Choose k centroids $c_1, ..., c_k$ at random.
- Choose the centroid that is geographically nearest to each data point x_i.
- Get the new centroids c_j by taking an average of all the data points in the jth cluster.
- When the difference between $L(X)$ on successive iterations is less than a predetermined threshold, the algorithm has converged. If this has not happened yet, repeat steps 2 and 3.

When applied to massive datasets, the k-means clustering algorithm's simplicity and effectiveness shine. But there are a couple of things to watch out for:

- You need to be careful when choosing k since it is a fixed parameter in the algorithm.
- Normalizing your data is a prerequisite to using k-means. It is common practice to normalize the jth coordinate x_{ij} to $(x_{ij} - \mu_j)/\sigma_j$, where μ_j represents the mean of the jth coordinates and σ_j denotes the standard deviation.
- Even if the features are numerically representable, you should not use k-means on them. Additionally, exercise extreme caution when using the k-means algorithm with a combination of numerical and binary features.
- The k-means algorithm is based on the spherical (globular) cluster assumption.

The diagrammatic representation of working of k-Means Clustering is shown in Fig. 8.13.

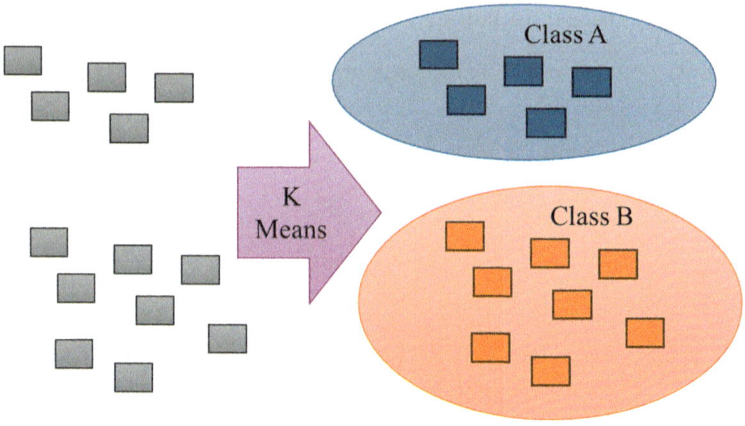

Fig. 8.13 Diagrammatic representation of working of k-means clustering

8.5.2 *Hierarchical Clustering*

Whereas the k-means algorithm relies on an operator-specified value k, hierarchical clustering methods do not. A challenging task that can significantly affect clustering results is choosing an appropriate k. The process of creating clusters using agglomerative (bottom-up) hierarchical clustering looks like this:

- Put each piece of data into its own distinct cluster.
- Join the two groups that are close together, with "most similar" being defined by some measure of distance like the Mahalanobis or Euclidean distance.
- Continue with step 2 until you've used up all of the clusters.
- Find the best clustering result by navigating the layers of the dendrogram as shown in Fig. 8.14.

One variant of hierarchical clustering that operates in the inverse direction is divisive hierarchical clustering. To reduce the number of clusters needed, we can begin with one that contains all of the data points and then partition it according to the distance metric. We check that each piece of data is in its own unique cluster.

8.5.3 *DBSCAN*

DBSCAN is a highly regarded clustering algorithm that is widely used and well-regarded for its strong performance in a variety of situations. The number of clusters is determined by the data itself, rather than being specified by the operator, which sets it apart from k-means. DBSCAN is a density-based algorithm that partitions datasets into subgroups based on high-density regions, in contrast to hierarchical clustering. This algorithm utilizes the following terminology:

- The radius ε is used to specify the range around a given point where the search for neighbors is conducted.

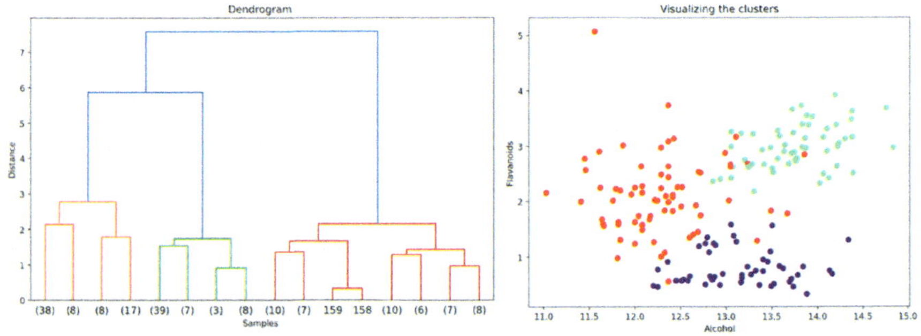

Fig. 8.14 An example hierarchical clustering

- The minimum number of points required to form a cluster is referred to as *minPoints*.

According to the given parameters, every data point is categorized into one of three groups: core, border, or noise.

- Points that meet the minimum point requirement within a given radius are referred to as core points.
- Border points are not considered as independent core points, rather they exist within the radius of certain core points.
- Noise points do not fall into the category of core or boundary points.

During a standard implementation, the clustering process entails examining each point in the dataset, calculating its distance to every other point, and subsequently assigning each point to its neighboring points based on a specific distance threshold. With this information at your disposal, you can easily categorize all the points as core, border, or noise. Once all the data points in the dataset have been classified, the DBSCAN algorithm proceeds with the following steps:

1. Select a point P at random from the set of unvisited points.
2. Mark P as visited and proceed if it is not a crucial point.
3. In order to create a cluster around a core point, it is necessary to recursively identify and incorporate all other points within the specified radius. This process also includes considering any points within the radius of other core points that have already been incorporated into the cluster. Suppose there exists a central point Q within the ε-radius of P. Q will be included in the cluster that is formed around P, along with all of its border points. If there is another core point R that falls within the ε-radius of Q, then the cluster formed around P will include core point R and all of its border points. This step is iterated recursively until all remaining core points have been taken over.
4. Keep iterating until every point in the dataset has been visited.

Although DBSCAN has proven to be effective in many scenarios, it does have certain limitations:

- DBSCAN's performance is affected when there are variations in cluster densities within the dataset. Choosing values for *minPoints* that are appropriate for all clusters in the data can be a challenging task. Utilizing the concept of ordering points to analyze and establish the clustering structure.
- Thoughtful parameter and minPoint selection is crucial for the algorithm's performance. Choosing appropriate values for these can be challenging without a solid understanding of the data's distribution and densities.
- The algorithm is non-deterministic, and the results can vary depending on the order in which points are visited during the random selection in step 1.

The diagrammatic representation of working of DBSCAN clustering is shown in Fig. 8.15.

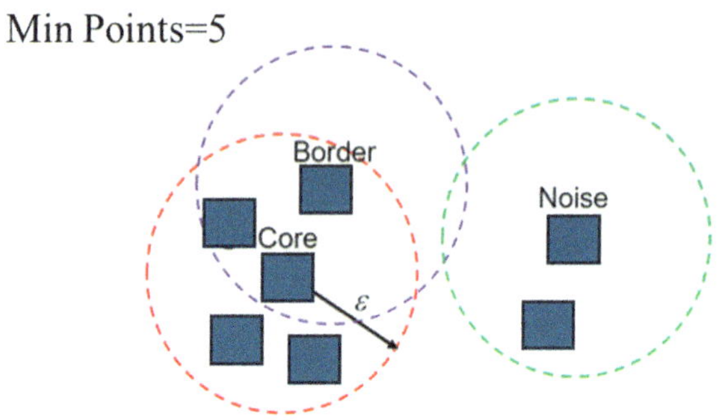

Fig. 8.15 Diagrammatic representation of working of DBSCAN clustering

8.6 AI/ML Implementation in Hardware Platforms

Present day intelligent applications are associated with AI/ML techniques. A lot of resources, money and effort is vested in generating an AI/ML based application. This starts from getting right data sets, training the data, developing the perfect model by refining it innumerable number of times and finally deploying it. Post deployment, monitoring is needed and even updating of the application is needed with time.

In the current embedded era, such AI/ML applications are deployed in end user devices. Such end user devices are generally in the form of system on chips (SoCs), where the AI/ML techniques need to be implemented.

Performing training and refinement in hardware platforms is too costly. Hence, AI/ML model training and refinement is performed in software platforms. Once a final model is generated, then it is deployed in hardware platforms (Fahim et al. 2021).

The diagrammatic workflow is shown in Fig. 8.16.

As discussed previously, some portions of the AI/ML model needs to be constantly updated, while some may be fixed. Present day SoCs comprise of several types of components like FPGAs, ASICs and even a softcore processor. These can

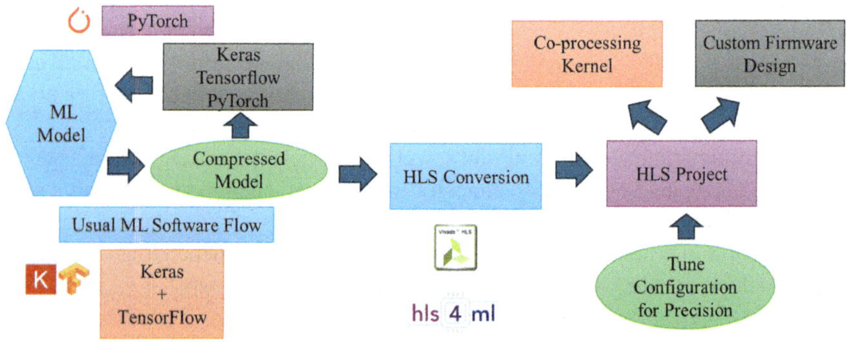

Fig. 8.16 Diagrammatic representation of workflow

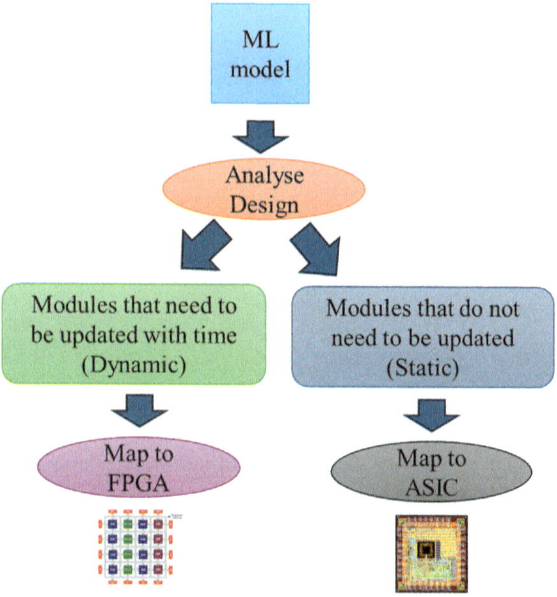

Fig. 8.17 Deployment logic

communicate among themselves. The system controller that handles all the real time tasks is generally implemented in the softcore processor. The portions that are fixed are implemented in ASIC platforms, while those that needs to be constantly updated is implemented in FPGA platform. This is illustrated in Fig. 8.17.

8.6.1 Deployment of AI/ML Model in FPGA Platforms

Implementing the final AI/ML model in FPGA platform is beneficial due to its property of dynamic partial reconfiguration. However, FPGA platforms are associated with limited number of resources and implementing the complete model may not be possible. Other than this, the processing rate is comparatively slower in FPGAs than traditional ASIC platforms.

Hence, it is wise to figure out the portions that need to be updated with time and implement that in FPGA platform.

For deployment, the design is initially coded with a hardware description language (HDL), i.e. VHDL or Verilog. Then this is synthesised and simulated in the FPGA based simulators to check any type of RTL errors. After this, it is passed thorough the place, map and route tools to map the design for a specific family of FPGAs. This leads to the generation of a bitstream, which can be dumped on the FPGA, as shown in Fig. 8.18. The bitstream configures the FPGA fabric and performs the AI/ML application.

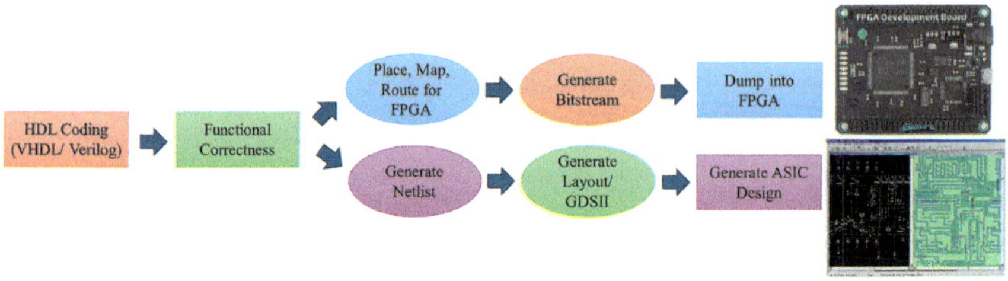

Fig. 8.18 Deployment flow on FPGA and ASIC

8.6.2 Deployment of AI/ML Model in ASIC Platform

ASIC platforms provide high speed for execution. However, these cannot be reconfigured with time or the implemented designs are rigid or cannot be changed.

Those portions of the AI/ML design are implemented in ASIC platforms, which need not be changed.

Like implementation in an FPGA, the initial phase comprises code writing in VHDL or Verilog, i.e. an HDL language. Then it is synthesized and a netlist file is generated. This netlist is used to generate the layout or GDSII file. This is then exported to fab to get the ASIC design. This is depicted in Fig. 8.18.

8.7 Implementation Strategies on FPGA Platforms

As discussed previously, implementation of AI/ML applications directly on hardware is definitly beneficial as it provides high speed in processing. However, the mechanism is quite costly and due to limited resources in an FPGA platform, it is not quite feasible to deploy an entire AI/ML application in the FPGA.

8.7.1 Objectives

Thus, implementation strategies on FPGA platforms needs to be associated with the following objectives (also depicted in Fig. 8.19).

8.7.1.1 Resource Utilisation

The main aim is to optimise the design of the neural architectures or AI/ML models and fit or map in the limited resources present in the FPGA.

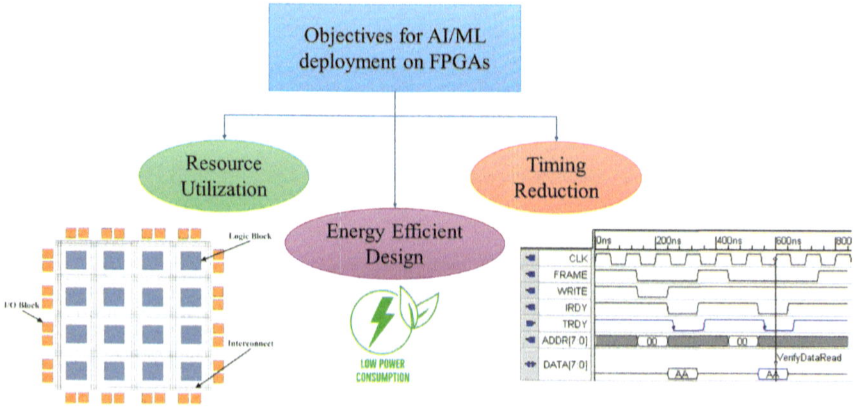

Fig. 8.19 Objectives of deployment of AI/ML applications on FPGAs

8.7.1.2 Energy Efficient Designing

Power consumption is a factor that needs to be considered. This indirectly affects the energy efficiency of the system. Hence, it is necessary to devise efficient mapping techniques so that the total power consumption of the system can be reduced and the system can be made more energy efficient.

8.7.1.3 Timing Reduction

For real time systems, timing is an important factor. Thus, developing techniques that can reduce the processing time is necessary. However, accuracy is an important feature that needs to be taken care of, when optimising the design to reduce the timing. More optimisation is associated with less accuracy. Hence, it is imperative to develop a strategy that not only reduces the timing, but also ensures very low loss in accuracy.

8.7.2 Potential Design Strategies

The potential mechanisms are (also depicted in Fig. 8.20):

8.7.2.1 Pruning

In this mechanism, some parts of the design is removed to optimise the entire design. This reduces the resource consumption directly. Indirectly, it reduces the power and

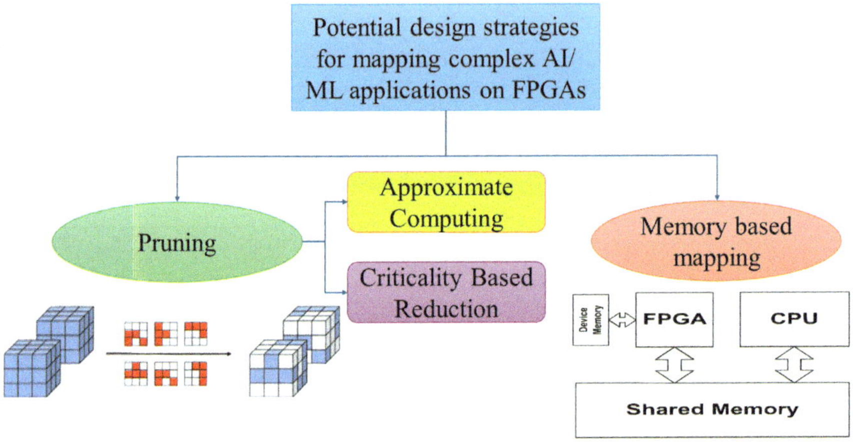

Fig. 8.20 Potential design strategies of AI/ML applications on FPGAs

timing based on the portion of hardware it utilises. Howver, the problem lies in loss of accuracy of results. Even quality of service (QoS) is compromised, as QoS is directly associated with refinement and execution of all types of functionalities, be it high critical or low critical.

Pruning can be done in various ways. Some are as discussed:

Approximate Computation: Many applications does not need full proof accuracy like video processing. Approximate results may be fine. Hence, some parts can be reduced. For instance, the number of iterations in a loop that is associated with perfecting an image can be reduced. The quality of final result depends on the number of times the loop is executed. To reduce power and time, the number of iterations may be reduced, which will give an approximate result.

Criticality based Reduction: For mixed critical systems, there may be some parts with high criticality and some parts with low criticality. Those associated with low criticality may be used very less and may not contribute to the generation of final result. These parts may be omitted to reduce resource consumption. For real time systems that is associated with mixed critical tasks, low critical tasks may be preempted, if some high critical task arrives randomly. Thus, no dedicated resource needs to be reserved for execution of high critical tasks that may arrive at random.

8.7.2.2 Memory Based Mapping

Present day FPGAs are associated with embedded memory blocks (EMBs). The number of EMBs is being increased by designers with aggressive technology scaling for FPGAs. Moreover, such advancement of technology has enabled designers to efficiently reduce the access energy of these EMBs in FPGAs. For instance, 53% reduction has been achieved by designers when a shift was made from Stratix II to

Stratix III and 48% reduction was made when a shift was made from Stratix III to Stratix IV. These EMBs are used by many applications to store inputs and intermediate data. But a large portion of the EMBs remain unutilized. These unutilized EMBs can be used for mapping computation intensive functionalities that utilize a lot of resources, are time consuming and dissipates a lot of power (Ghosh et al 2013).

The ideal scenario is to map each and every function to the EMBs. However, the number of EMBs is not sufficient. Hence, it is wise to order the computational intensive functions as per their resource consumption or power dissipation or time consumption (as per the objective). Some functions may also be fused, if they are in a pipeline. Then the functions in the top of the list is mapped first and the others are followed as per the availability of the EMBs.

This reduces resource utilisation, power consumption and time. However, the main issue lies in the size of the EMBs, as it cannot support all the functionalities of an AI/ML application. Moreover, such a process makes the system quite rigid, as in case of a change, entire contents of the EMBs need to be re-written (Guha et al 2023).

8.8 Secrets and Security

Significant effort, time, resources and money is vested in developing an AI/ML model. Rather than developing a custom neural network model (if needed), generation of weights and biases is quite important. If an adversary gets hold of these, then it is possible for the adversary to duplicate the model and sell it at low price or even serve users at low cost. Moreover, if a new custom model is not even generated, at least the weights and biases vary as per the application. If these weights and biases are stolen by an adversary, then the adversary can easily generate a duplicate and earn money (Guha and Bhattacharyya 2023).

For the present scenario, the weights and biases are the secrets, which needs to be protected from adversaries. Additionally, if a custom neural architecture is created, that also needs protection from attacks of an adversary.

8.8.1 Attacks

Adversaries may try to get acquainted with the custom generated neural model or the weights and biases that are generated after a lot of training activities. The weights are biases are generally stored in a memory, which may be on chip or off chip. In either case, these need to be loaded in the processor during execution. They may either be stolen during transit or they may be tampered that will result in erroneous result generation. Even they can be delayed that can result in denial of service. While stealing them is passive attack, modifying or tampering them or causing denial of service are active attacks.

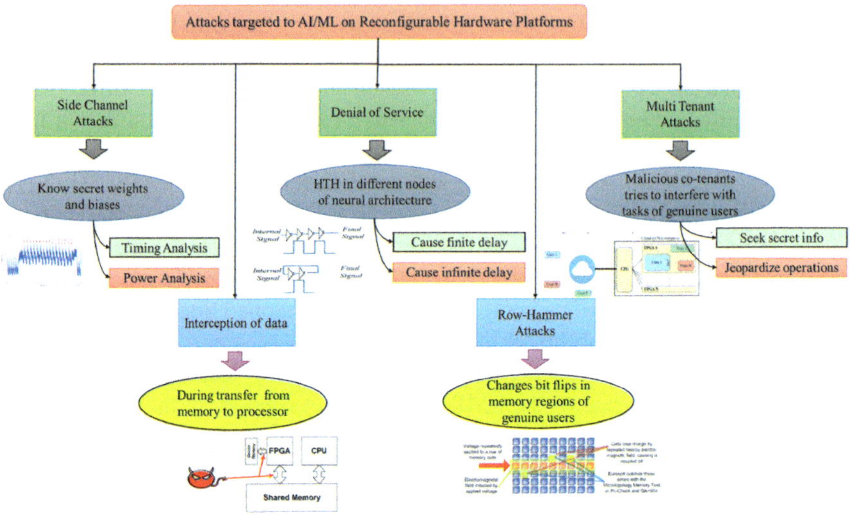

Fig. 8.21 Attacks for AI/ML applications on FPGAs

Various types of attacks are possible. Some of these are (also depicted in Fig. 8.21).

8.8.1.1 Side Channel Attacks

The adversary may try to know the weights via intrinsic side channel observations or leakages, such as the timing of operations or via delay or via power analysis, etc.

For instance, different functionalities of a composite system takes different times to execute and even dissipates different amounts of power. These can be noted for different systems and via reverse analysis, adversaries tries to find out the various operations.

Moreover, when different weights or bias operates, then the timing of operation as well as the power dissipation varies. If an application uses a generic neural network architecture but its weights and biases will be unique based on the training and datasets used. In such scenarios, the neural architecture is known but the weights and biases are secrets. Based on the timing analysis or power analysis, adversaries can get acquainted with the secret weights and biases.

8.8.1.2 Interception

The secret weights and biases are stored either on-chip memory or off-chip memory. In either case, these needs to be transported to the processor. These can be intercepted during transit by the adversary. This procedure is quite easy when the weights and biases are stored in off-chip memory and the procedure may be as simple as tapping the connecting wires or monitoring traffic to and from the memory. The process becomes a bit complex when the weights and biases are stored in an on-chip memory and adversaries generally deploy hardware trojans to analyse the traffic and reveal the data via side channel parameters.

Reconfigurable hardware or FPGAs are associated with execution of different applications at different instance of times, with the aid of their property of dynamic partial reconfiguration. Due to low resources in FPGA platforms, the weights and biases of the different applications are generally stored in off-chip memories and transferred dynamically to the FPGA processor. These can be intercepted during transit by adversaries.

8.8.1.3 Denial of Service Attacks

Hardware Trojans may be present in the FPGA fabric or in the bitstreams which may delay execution. This will prevent the intelligent neural network based applications to complete their working before the deadline. These will affect real time systems and may cause fatal consequences.

Such hardware trojans are essentially delay inducing circuits which may contain buffers or buffers within a loop. The delay can be either finite or infinite. Though finite delays can be compensated with increase in frequency of operations at later stages. But if the delay is infinite, then the module needs to be replaced. In either case, the module associated with such denial of service needs to be found and compensative actions have to be taken.

8.8.1.4 Multi Tenant Attacks

Reconfigurable hardware features dynamic partial reconfiguration, with the aid of which spatio temporal scheduling of tasks can be performed in real time. Spatio temporal scheduling means scheduling of real time tasks in the FPGA with respect to space (spatio) and time (temporal). For this, the FPGA fabric can be partitioned into various virtual partitions or portions (VPs) and different tasks can be scheduled in these VPs at different instance of times. These tasks are procured from different users or tenants.

Thus, an FPGA fabric can be partitioned into different VPs, which can be allocated to different tenants to execute their tasks at different instance of times. So, an FPGA can serve multiple tenants.

However, some tenants can be adversaries and may try to jeopardise the operations of genuine tenants. Multi tenant attacks can jeopardise the CIA (confidentiality-integrity-availability) of the system. For instance, malicious co-users or tenants may try to intercept the transmission to snoop into the secret weights and biases that are being transferred, which may affect system confidentiality. Some malicious tenants may execute the tasks in the FPGA VPs for significant amount of time and more than the time interval given to them forcing other genuine users to wait or intentionally delaying their execution. This will affect system availability. Moreover, bitstreams that are stored in memory can be modified by malicious co-tenants that will affect system integrity.

8.8.1.5 Row Hammer Attacks

Row hammer is a phenomenon where a tenant may access the memory locations in such a way that will cause bit flips in the neighbouring memory locations (Mutlu and Kim 2019).

Problems will arise where the memory is divided into multiple slots and each slot is denoted to a tenant for storing its data. Now, a malicious co-tenant may access its boundary memory cells in such a frequency that will cause bit flips in the memory cells of the memory rows that are present in another user's slot. Such bit flips will cause change in the data stored and will generate erroneous results (Guha and Chakrabarti 2022).

If the memory cells that encloses the weights and biases are changed, then the AI/ML application will generate erroneous results.

Row hammer attacks is a special type of multi tenant attacks, where the malicious co-tenant causes an attack intentionally, but remaining in its legal boundaries.

8.8.2 Solutions

Attacks and mitigation strategies go hand on hand. Suitable solutions also exist for the different types of attacks. As reconfigurable hardware is different from normal hardware for fabric partitioning, bitstream deployment and execution, hence, the security mechanisms is also a bit different.

These are described below (also illustrated in Fig. 8.22).

8.8.2.1 Logic Locking

In logic locking, the secret functionalities are locked, i.e. the functions can only operate if a series of secret bits are entered. This is generally done by the designers. Adversaries who tries to intercept the design cannot use it as they do not know the secret bits, which serves as a lock.

For AI/ML applications, essential functionalities encompass the operations that utilise the weights and biases. These functions are locked via secret bits. If the user enters the secret bits, then only the functionalities are executed and results are generated.

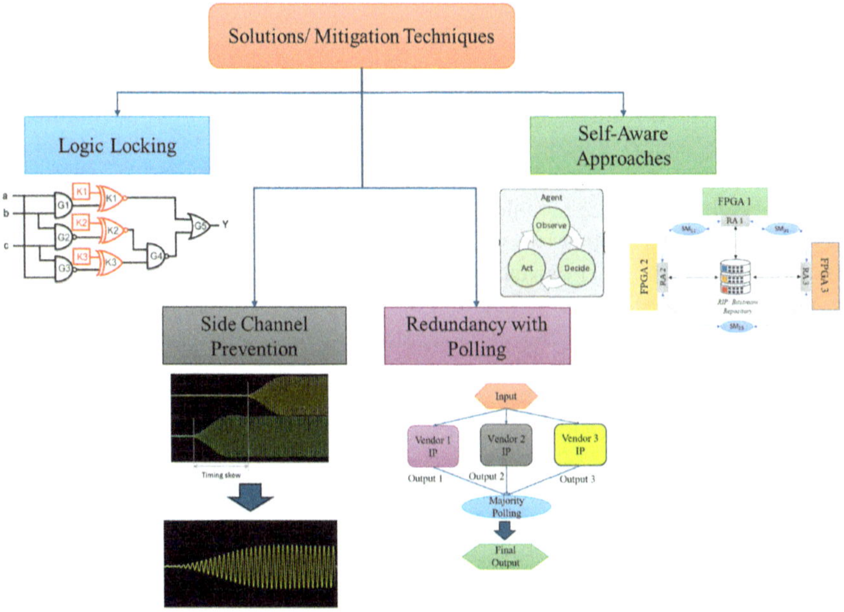

Fig. 8.22 Solutions/Mitigation techniques for AI/ML applications on FPGAs

8.8.2.2 Side Channel Prevention

Preventing side channel attacks is essential and important in the arena of ensuring security. Side channel analysis is done based on analysing the difference of side channel values. This can be done via confusion or synchronisation. These prevents leakage of internal structure, as well as secret information to adversaries.

In confusion, the designer adds random logic in the various modules of a design. This randomises the timing or power dissipation. The timing and power varies in each iteration (based on the random number generated at that instant of time). This prevents an adversary to reverse engineer and get acquainted with the internal structure and the secret parameters.

The other method is synchronisation, where the designer implants an additional module which synchronises the timing of output generation (Guha et al 2017a, 2015). Such a module compromises of two units, one is the monitoring unit that notes the amount of time taken for operation of each unit and the other is the synchronising unit, which aligns the output timing. The same mechanism can be associated with power.

8.8.2.3 Redundancy with Polling

This mechanism is associated with critical functionalities, which needs high accuracy in results. The bitstream of a particular operation is procured from three different vendors and executed in three different virtual portions of an FPGA. Then majority polling of the results is done, before generating the output.

In general, all the three modules cannot undergo fault at the same instance. Only one may be faulty, the other two will generate perfect results. The one whose result mismatches with the other two can be faulty. The fault can be either in the bitstream or in the FPGA VP fabric. For fault diagnosis, another bitstream that generated the correct result is executed in the VP associated with the faulty operation. If the new result generated is correct, then fault was associated with the bitstream, else the fault was associated in the FPGA VP.

With increase in the number of modules, i.e. from three to five or seven, reliability can be increased. But this will increase the overhead in area and power. Hence, the standard is to use three modular redundancy with majority polling.

8.8.2.4 Self Aware Approaches

Redundancy based majority polling induces high overhead in area and power, though ensures accuracy immensely. However, high overhead is unwanted. The overhead problem is of significant concern for AI/ML application. This is because most of the layers of operation are associated with computations with weights and biases. Triple modular redundancy for each layer of operation will definitely enhance the area and power overhead for the entire system.

Hence, a lightweight solution is needed. For this, self aware modules are created. These self aware modules are plug and play entities. These monitor the system based on their side channel parameters and on determining any change, triggers the fault diagnosis unit. Thus, the fault diagnosis modules, which works similar to the triple modular mechanism is only utilised when an undesired scenario is observed (Guha et al 2018).

Such self aware modules can be associated with added functionalities. For instance, in a distributed system platform, where multiple resources are present with multiple self aware modules, the self aware modules can communicate with each other via shared memories. On detecting a vulnerability, the self aware module can outsource the task to another self aware module, which can accelerate the task execution and complete within the required deadline. This is particularly important for real time systems, associated with AI/ML operations with deadlines.

Questions

1. Justify the need of AI/ML in cybersecurity.
2. Describe the various components of a ML based sytem and also state how these can be attacked by an adversary.
3. State the potential difference between classification carried out by (i) Naive Bayes (ii) Logistic Regression (iii) SVM.
4. State how the mechanism of clustering varies for the following: (i) k-means (ii) Hierarchical (iii) DBSCAN.
5. State the potential mechanism (including the various platforms that may be necessary) to implement an AI/ML model in a hardware platform.
6. State the various steps for implementing a design (starting from coding in HDL) to implement on (i) FPGA (ii) ASIC.
7. State how a designer can decide which portions in a complex design must be implemented in FPGA and which portions must be implemented in ASIC.
8. What are the key objectives for implementation of an AI/ML model in FPGA platform?
9. State and describe the various types of attacks that are possible for an AI/ML model in reconfigurable hardware platforms.
10. State the various mitigation techniques for related threats on AI/ML model implemented in reconfigurable hardware platforms.

References

Fahim F, Hawks B, Herwig C, Hirschauer J, Jindariani S, Tran N, Carloni LP, Di Guglielmo G, Harris P, Krupa J et al (2021) An open-source codesign workflow to empower scientific low-power machine learning devices. arXiv:2103.05579

Gama J, Brazdil P (1995) Characterization of classification algorithms. In: Portuguese conference on artificial intelligence. Springer, pp 189–200

Geluvaraj B, Satwik P, Ashok Kumar T (2019) The future of cybersecurity: major role of artificial intelligence, machine learning, and deep learning in cyberspace. In: International conference on computer networks and communication technologies: ICCNCT 2018, Springer, pp 739–747

Ghosh A, Paul S, Park J, Bhunia S (2013) Improving energy efficiency in FPGA through judicious mapping of computation to embedded memory blocks. IEEE Trans Very Large Scale Int (VLSI) Syst 22(6):1314–1327

Guha K, Saha D, Chakrabarti A (2017) Real-time SoC security against passive threats using crypsis behavior of geckos. ACM J Emerg Technol Comput Syst (JETC) 13(3):1–26

Guha K, Bhattacharyya G (2023) A self aware security approach for real time neural network applications from row hammer attacks in multi FPGA multi user environment. In: 2023 IEEE international symposium on defect and fault tolerance in VLSI and nanotechnology systems (DFT). IEEE, pp 1–4

Guha K, Chakrabarti A (2022) Criticality based reliability from rowhammer attacks in multi-user-multi-FPGA platform. In: 2022 35th international conference on VLSI design and 2022 21st international conference on embedded systems (VLSID). IEEE, pp 234–239

Guha K, Saha D, Chakrabarti A (2015) RTNA: Securing SOC architectures from confidentiality attacks at runtime using ART1 neural networks. In: 2015 19th international symposium on VLSI design and test, pp 1–6. https://doi.org/10.1109/ISVDAT.2015.7208048
Guha K, Saha D, Chakrabarti A (2018) SARP: self aware runtime protection against integrity attacks of hardware trojans. In: International symposium on VLSI design and test. Springer, pp 198–209
Guha K, Trivedi AR, Bhunia S (2023) Energy efficient memory-based inference of lstm by exploiting FPGA overlay. In: 2023 international joint conference on neural networks (IJCNN). IEEE, pp 1–7
Jung A (2022) Machine learning: the basics. Springer Nature
Mutlu O, Kim JS (2019) Rowhammer: a retrospective. IEEE Trans Comput-Aided Des Int Circuits Syst 39(8):1555–1571
Xu R, Wunsch D (2005) Survey of clustering algorithms. IEEE Trans Neural Netw 16(3):645–678

Chapter 9
Digital Forensics

Objectives of The Chapter

- Understanding Cybercrime.
- Introducing Digital Forensics.
- Phases of Digital Forensics Investigation.
- Guidelines and Principles for Forensic Evidence.

9.1 Cybercrime

In today's digital age, the ever-expanding realm of computer technologies and the Internet has provided criminals with new opportunities to engage in illicit activities (Holt and Bossler 2015).

Cybercrime refers to the utilisation of the Internet, computers, and other technology for the purpose of engaging in criminal activities. It encompasses both technologically specific offences that are only feasible via the use of computer technology, as well as traditional crimes facilitated by a computer. The advent of cybercrime has led to an increased range of criminal acts, with technology and the Internet giving rise to numerous new forms of crime.

With the advancement of technology, we've seen the rise of different types of software and network threats. These include hacking, which involves unauthorized access to computers, and malware, which refers to malicious software like computer viruses, worms, and Trojan horses.

Previously, there was a prevailing belief that hardware systems have a sufficient level of reliability. These are immune to software threats. Consequently, designers extensively utilised and implemented them. However, the emergence of hardware

K. Guha et al., *Cybersecurity for Reconfigurable Hardware Based Critical Infrastructures*,
https://doi.org/10.1007/978-3-031-67591-1_9

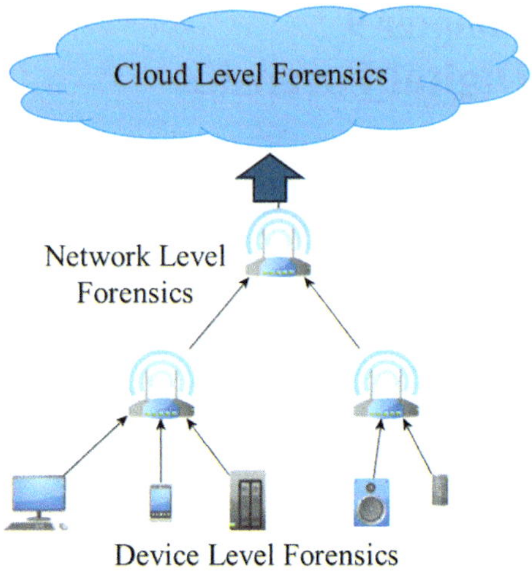

Fig. 9.1 Workflow in digital forensic process

trojans, counterfeit intellectual properties (IPs), IP theft, forgeries, and other related issues have undermined the integrity of the hardware root of trust. These factors can impact all aspects of the CIA triad. For instance, they can compromise confidentiality by disclosing classified information, compromise integrity by producing inaccurate outcomes, and compromise availability by creating a denial of service.

Present day technologies utilize resources not only in the device level, but also in the edge and cloud level. Hence, the range of threat increases and it can be concluded that digital forensics must range throughput the cloud to thing continuum (Hou et al. 2019). This is depicted in Fig. 9.1.

To what extent a crime involves the use of a computer is what defines it as a cybercrime. Cybercrime can only be defined as an offense where computers play an integral role, either as the intended victim's device or as the means of perpetration.

9.1.1 The System as Target

A adversary may target the server or system resources, where the adversary:

- attempts to breach a system and/or
- intrudes upon the system and pilfers confidential data
- bombards the system with external attacks, or
- inflicts damage upon the computer

Illustrative instances encompass a range of activities such as hacking, decryption, distributed denial of service (DDoS) attacks, hacking, and malware distribution.

This is addressed in a broad sense (without focusing specifically on hardware-based reconfigurable system architectures):

9.1.1.1 Cracking and Hacking

An individual who endeavours to acquire unauthorised access to a computer system is known as a hacker. Hacking tools are readily accessible on the internet, enabling individuals with minimal computer expertise to acquire the means to infiltrate a computer without authorization.

Cracking is the act of an assailant attempting to obtain "unauthorised access to a computer system with the intention of committing another crime, such as destroying information contained in that system".

9.1.1.2 Distributed Denial of Service (DDoS) and Denial of Service (DoS)

Such attacks are an endeavour to impede the functionality of a particular service for its consumers. When a network server is overloaded with (false) authentication requests, it is generally unable to process them. The assault overwhelms the resources of the targeted computer or computers, resulting in their inability to grant server access to other computers that are submitting valid requests. Eventually, the server is turned down. When multiple servers are targeted in a planned manner, then such an attack is termed as a DDoS attack.

9.1.1.3 Dissemination of Malicious Software

Malware, also known as malicious software, has the ability to corrupt a computer system or gain unauthorized access to steal valuable information. Malware encompasses computer viruses, worms, Trojan horses, spyware, and botnets.

Virus on the computer: A computer virus is a malicious software programme that is specifically engineered to infect and disrupt a computer by means of interfering with communications or excessively taxing its resources before it can be eliminated. In some cases, a computer virus may be intentionally created to cause inconvenience to a user by means of displaying a message at a predetermined time on the user's screen. Computer viruses are designed to adsorb (or piggyback) on programmes or files. They are transmitted through the action of a computer user, such as opening an infected email attachment. Computer worms and viruses are both types of computer worms. In contrast to computer viruses, computer worms do not depend on specific files or programmes for replication. Unlike computer viruses, worms replicate and propagate without requiring user interaction.

Trojan Horse: Trojan horses download and install hidden programs while making the computer user think they are installing real software. People who use computers download and install software, and these hidden tasks are done at the same time. By deleting files from a user's system, these features could harm a computer or steal sensitive data.

It is also possible for Trojan horses to add logic bombs to computer systems. A logic bomb can't make copies of itself. When something happens, like a set date and time or running a certain program a certain number of times, it awakens a program that had been sleeping. It is possible for a logic bomb put into a user's computer to delete the hard drive as soon as the system starts up again.

Spyware: Spyware is computer software that lets someone watch what a computer user does or sees on a computer that has this software installed. Spies can also get information about users (like credit card numbers and passwords) without their knowledge and give it to people who are interested. One type of this software is a keylogger, which, once installed, keeps track of every key the user presses and sends that data back to its source. Sniffers are programs that look at and record network traffic. They can also be used to get usernames, passwords, and other information that can be used to find out who someone is.

Botnets: "Bot code", which is malicious code, can be used to take over someone else's computer (or computers). Bot code lets people remotely take control of computers that have it without the users knowing. It is the people who control botnets. A botnet is a group of zombie computers. Individuals who infect computers with bots may collect data from those computers, which may include personal details, credit card numbers, email account information, usernames, and passwords; the botnet may also do whatever the individual who infected it wants it to do.

9.1.2 The System as a Tool for Crime

Criminals use computers to do this kind of cybercrime. Theft of intellectual property, phishing, cyberharassment, cyberstalking, cyberbullying, illegal online sales of controlled substances and prescription drugs, and Internet gambling are all examples of this type of crime.

These things have nothing to do with reconfigurable hardware-based systems, but it's still good to know about them. Some of these are described below and also depicted in Fig. 9.2.

9.1.2.1 Infringement on Copyright

Copyright law says that the person who made an original work has a limited right to make copies or distribute those copies. Pirated movies, TV shows, music, software, books, and video games are examples of copyright infringement. It happens when someone copies all or part of someone else's work without permission or when they let other people copy their work without their permission.

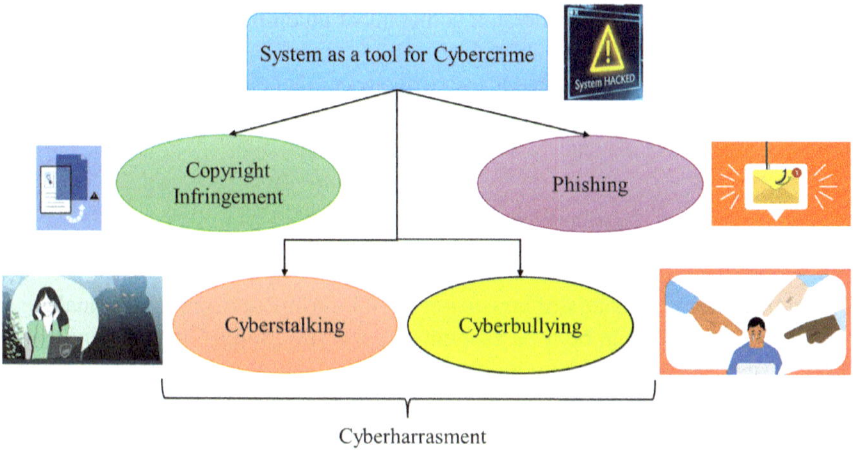

Fig. 9.2 System as tool of cybercrime

With the evolution of embedded devices, the intellectual property rights extends to such hardware oriented devices.

9.1.2.2 Phishing

Cybercriminals can use computers to do bad things like phishing. People pull off phishing by pretending to be real businesses in order to get valuable personal information from people who use communication tools, like account and credit card information. The majority of the time, cybercriminals will pose as legitimate banks in an email and ask their victims to verify their information by clicking on a link. Clicking on the link will lead you to a website that mimics the appearance of the bank's. If the victim were to enter sensitive information and then click "Submit," the perpetrator would immediately have access to that data. After that, the user is redirected to the bank's official website, where they are asked to enter their login credentials once more.

9.1.2.3 Cyberharrasment

Assaulting, bothering, or otherwise annoying someone on purpose over the Internet, email, or other forms of communication is considered cyberharassment. As well as other things, it can include rude and offensive comments about gender, race, sexual orientation, nationality, or religion.

Cyberstalking is when someone repeatedly bothers or threatens someone over the Internet, email, or other electronic communication devices. Someone who is cyberstalking may bother or threaten someone so much that the target (the person being stalked) worries about their safety.

Cyberbullying is when someone uses a phone or other electronic device to bother, insult, or embarrass a child, teen, or adult.

9.2 Introducing Digital Forensic

Digital forensics is the study of how to get digital information, process it, analyze it, and store it so that it can be used as evidence in civil, criminal, and administrative cases (Årnes 2017). Within the field of forensic science, digital forensics studies how criminal law, procedure, and evidence work on computers and other electronic devices.

You can get this kind of information from computers and other electronics like printers, scanners, copiers, CDs, DVDs, Blu-ray discs, external hard drives, USB flash drives, magnetic tape data storage devices, cameras, cell phones, landline phones, fax machines, PDAs, portable media players, and game consoles. You can get back data from files that already exist, even ones that have been deleted, encrypted, or damaged, or you can watch what users are doing in real time. A lot of different types of crimes, including cybercrimes and computer misuse, can use data from computers and other electronic devices as proof. It can also help catch and prosecute criminals, stop illegal activity from happening again, look into employee misconduct, and fire workers.

This field of forensics encompasses not only computer forensics but also includes mobile phone forensics, personal digital assistant (PDA) forensics, and network forensics. With the IoT digital age, hardware forensics has also gained importance. Especially, with the advent and widespread use of reconfigurable hardware, it is necessary to determine who used the hardware and when.

In short, "*The legal use of computer science and investigation methods to look into digital evidence after the right search authority, chain of custody, mathematical validation, use of validated tools, repeatability, reporting, and possibly an expert presentation*" is termed as digital forensic by Ken Zatyo.

Digital forensics isn't just about computers like laptops and desktops. The field looks at things like mobile devices, networks, and "cloud" systems. It also includes analyzing pictures, videos, and sounds (both analog and digitally). This kind of analysis is usually about finding the truth, comparing things, and making things better. In the present era, such computing is directly done in hardware to escape software related threats (along with other advantages). However, computing in hardware is also not fully secure.

With the advancement of reconfigurable hardware and its deployment in various infrastructures from end user devices to cloud infrastructures, digital forensic techniques for reconfigurable hardware based infrastructures need to be identified. While the threats are similar to those discussed in the previous chapters, however, their identification strategy varies in different scenarios.

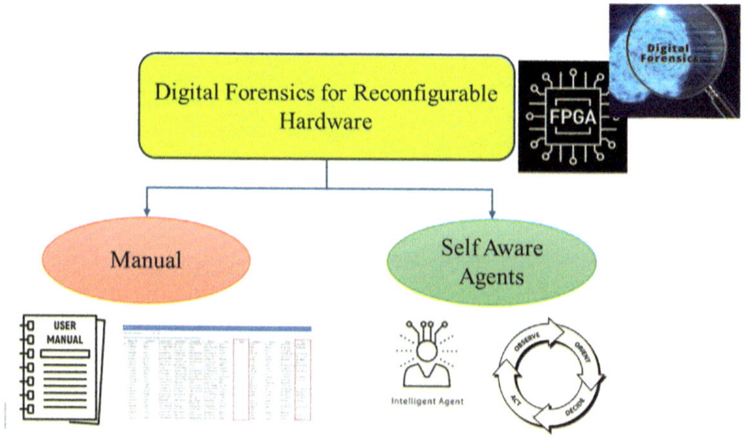

Fig. 9.3 Digital forensics for reconfigurable hardware

For reconfigurable hardware based platforms, which are generally headless, forensic investigation can be performed on the go or at the end of a certain phase. If the forensic is performed on the go, then the abnormality detected by the agents are communicated to the administrator via setting of alarms or via wireless channels. Else at the end of the phase, the administrator manually analyses the operations of the agent or controller and detects the malfunction.

This is depicted in Fig. 9.3.

9.3 Phases of Digital Forensics Investigation

The acquisition, identification, evaluation, and presentation of digital data constitutes the major steps in digital forensics investigations.

These are depicted in Fig. 9.4 and also described below:

9.3.1 Acquisition

The process of finding evidence, collecting it, and writing it down is all part of the acquisition step in digital forensic investigations. The digital forensics expert or, for a headless system, the module that is responsible for finding and fixing a vulnerability must write down the following parts of the computer search.

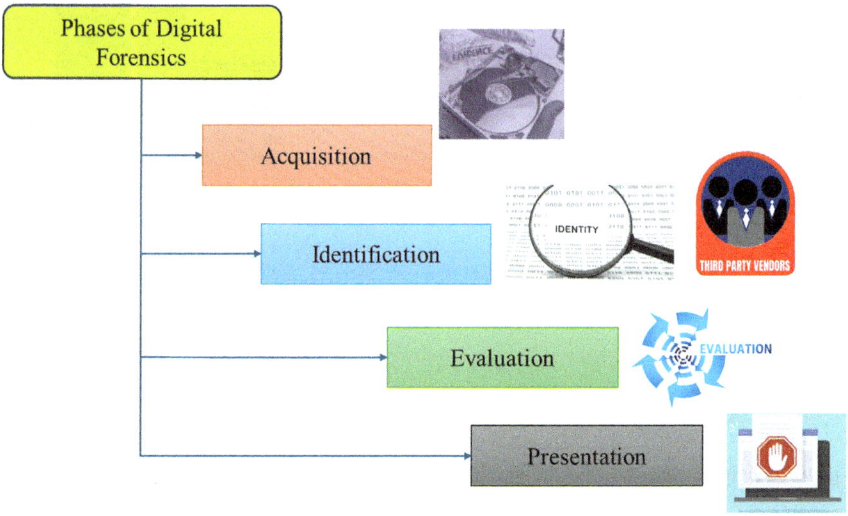

Fig. 9.4 Phases of digital forensics

- *Which evidence was obtained?*
  For reconfigurable hardware based systems, a particular malfunction once made is associated with an effect. This effect is the evidence that needs to be obtained or detected.
- *Who retrieved the evidence?*
  This is generally associated with the working of the agents or the controller of the system. For instance, there is an agent with each FPGA or a group of FPGAs that oversees their action and controls the real time tasks. This agent observes any malfunction and is associated with retrieving the evidence, which is the effect left after a malfunction.
- *Where was the evidence gathered?*
  Such effects are generally associated with abnormal behavior associated with side channel parameters like time, power, leakage current, etc. This may also be associated in the memory of resources like change in BRAM data of FPGAs.
- *When was the evidence gathered?*
  This is determined by the programming that is done on the controller or agents. Some are programmed to monitor at periodic phases, while some are done at the end. The best results come when the monitoring is performed continuously at runtime.

9.3.2 *Identification*

During this part of the investigation, known as "identification", the investigating body or module gives a good enough rationale and records where the evidence came from and why it matters. Given the potential for multiple interpretations of the evidence, this step sets out the circumstances surrounding its discovery. The evidence's physical and logical contexts are both taken into account.

For instance, if an operation is found anomalous by a self monitoring agent, then the agent tries to find out whether the malfunction was a result of the bitstream or was associated with a malfunction of the FPGA fabric. For this, it may execute another bitstream performing the same functionality but procured from another 3PIP vendor and execute it in the original FPGA. This is the mechanism of online investigation.

The investigator examines the locations of evidence within an operating system's file system which is responsible for cataloguing the names and locations of files on a storage device, such as a hard drive. A file on a hard disk is placed at a specific location and an entry of the file identifier is made in an index. The process of indexing and layout specifies what is called as a file system.

Data extraction is possible through the retrieval of file system information that discloses attributes such as file name, location, size, and more. Active file records, recovered deleted file records, and unallocated space all have the potential to yield data recovery outcomes. It is also possible to retrieve it from the file slack, which is the empty space at the beginning or end of a file system or between two logical units.

Making previously hidden evidence visible to the forensic analyst for review in the next stage of digital forensics—evaluation—is also part of the identification stage. For the sake of the process's credibility and the admissibility of pertinent evidence in legal proceedings, the investigator must meticulously document the physical and logical setting of the evidence.

9.3.3 *Evaluation*

As part of digital forensics, the evaluation step involves analyzing the data retrieved during the investigation to establish its importance and relevance to the current case.

- How does the evidence point to the anomaly, the accused, or the victim?
- When and how did the data become part of the system?
- Did the suspect or someone else place the data there?

The defendants, which in this present case are the third party vendors from whom the resources are procured, mostly claim that they were not responsible for the materials found on their computers as some malware programs might have downloaded these materials onto their computers. Based on the evidence gathered, investigators attempt to determine the "*who*", "*what*", "*when*", "*where*", "*why*", and "*how*" of the crime.

For the online investigation mechanism by a self aware module, discussed in the previous case. If abnormality is still detected in the redundant operation carried out by another bitstream procured from another third party vendor but performing the same functionality on the same FPGA, then it is concluded that the FPGA fabric is malicious, else the bitstream is malicious.

For the present instance, the self aware module may record it directly into the log file or writes it in a shared memory, which is read by the admin and recorded in the log. Physical media, like a hard drive, may store evidence in the form of log records. Consequently, methods like file carving and keyword searches can physically retrieve data from a hard disk. By analyzing the file's header, footer, and other identifiers, file carving can search a hard drive for particular files. This is how an investigator can search corrupt directories or damaged media for files or fragments of damaged or deleted files.

9.3.4 Presentation

Reporting pertinent data found during the investigation is what the last stage, presentation, is all about. Outside parties may be expected to evaluate evidence at this stage.

For distributed autonomous systems, the identification needs to be communicated to other agents of the system. For this, the agent may share this information directly with other agents via sending direct messages or via a shared memory communication. It can also happen that the agents reads the main log file that is updated by the administrator from time to time. However, for security concern, this has to be only a read only mode for the agents and only the administrator has the permission to write.

Based on the reports, a third party vendor may be blacklisted. If a vendor is blacklisted, then resources procured from the vendor may not be used in any future operations. This may lead to non-procurement of resources in future from the vendor. Temporary stoppage in usage of resources of a blacklisted vendor is common. However, if non-procurement of resources are done, then solid evidence is required.

This requires computer forensics specialists to be court-ready during the post-operational phases. It is common practice to have witnesses defend their credentials, methodology, procedures, evidence handling, and conclusions while testifying. The ability to articulate one's findings to various audiences is another essential skill for computer forensics experts. This includes juries, judges, attorneys, corporate executives, and government officials. While this is going on, there is still time to contest the evidence's chain of custody, a chronological record that shows when and who had the item in question.

9.4 Electronic Evidence

Any information pertinent to an intentionally performed abnormal operation that is detected and recorded by the monitoring agents or modules and runtime, and then duly investigated and verified by forensic experts in the post-operation phases is referred to as evidence.

The following reasons motivate the search for evidence:

- To establish that an actual malfunction occurred, which was created intentionally.
- To connect a specific person or third party vendor that is sending the resources for operations.
- To disprove or support the testimony of third party vendors that are supplying the resources for operation.
- To identify a suspect and to understand their motive.

Typically, electronic evidence includes any data that can be retrieved from a system or device and utilized to establish or refute a crime or policy infraction.

This kind of proof can prove both intent and possession. For example, if a person's computer contains images of children, it could be a sign of possession. By sorting the photos by the child's screen name, we could find out what they were up to.

Similarly for reconfigurable hardware based distributed system, if a vendor has attacked a particular component of a distributed system and tries to take control of that component, then it indicates possession. However, if the attack is made in such a way that it spreads to other components and jeopardizes the entire system, then its intent.

It is possible to offer an alibi or support a claim with electronic data. Computer forensics investigations seek to analyze evidence in order to determine when, how, why, and who was involved in an incident.

9.4.1 Evidence Types

There are various types of evidence that can be used to prove or disprove a piece of information. This section delves deeper into the testimonial, documentary, demonstrative, physical, direct, circumstantial, and hearsay evidence.

Some of these are depicted in Fig. 9.5 and also described below:

9.4.1.1 Testimonial Evidence

As a rule, evidence in court proceedings typically relies on the statements of eye-witnesses or experts who have testified under oath. To be considered competent, a witness must meet certain requirements, such as having first-hand experience with the matter at hand. Expertise in this area can come from direct observation of the

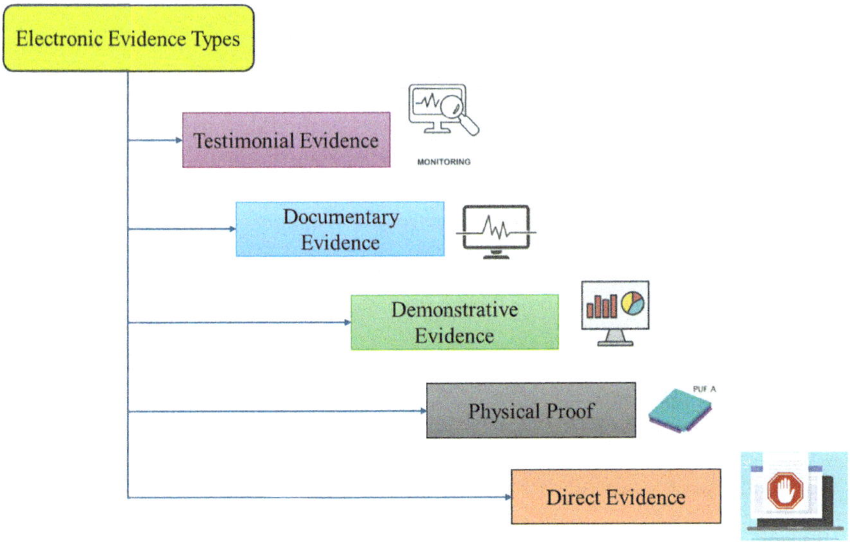

Fig. 9.5 Evidence Types

crime scene (in the case of an eyewitness) or from specialized study that allows one to confirm or deny claims made in the case. Individuals called to serve must possess this specific knowledge.

For reconfigurable hardware based systems, the agents associated with monitoring and security are programmed in such a manner that it can detect and provide first hand evidence.

9.4.1.2 Documentary Evidence

To be admissible as evidence in a court of law, any written, video, or audio recording must have its authenticity proven. Among these are computer printouts, manuals, and company records. Verification of the authenticity of all documentary evidence is essential. Although there are some notable exceptions, expert testimony is usually required to verify the validity of the evidence. Extraneous proof of authenticity is not necessary for some documents. These include official publications, certified copies of public records, and domestic public documents that are under seal.

The agents normally works based on observing several side channel parameters like power, delay, etc. Thus, it documents the deviation from original, which can lead to undesirable consequences.

9.4.1.3 Demonstrative Evidence

While not conclusive in and of itself, demonstrative evidence serves to (i) shed light on supporting evidence, (ii) paint a picture of what happened, or (iii) illustrate a scenario comparable to the one at hand.

As an example, visual aids such as diagrams, drawings, maps, models, and sketches are employed to enhance the comprehension of the evidence. Additional forms of evidence that present a visual representation of events at a crime scene or as recounted by a witness in court include photographs and videos. Videos can be utilized to support a witness's testimony as long as their content is thoroughly reviewed to establish credibility, applicability, and expert knowledge.

To be considered admissible in a legal proceeding, demonstrative evidence must typically fulfill the following conditions:

To demonstrate, the module essentially re-executes the same task with different set of resources and records both the cases. This aids in clear understanding of the malicious scenario and which resources were associated with that case.

9.4.1.4 Physical Proof

Extractions from hard drives and other electronic devices are not the only sources of evidence in a computer forensics investigation. It is possible to find tangible proof, often called real evidence. This evidence can back up the claims of the suspect, the witness, or the victim; link the suspect to the crime scene or the victim; or even disprove a suspect's guilt. Some examples of tangible evidence include weapons, knives, glass, and documents whose authenticity is still up for debate, such as handwritten or typewritten documents.

The resources that functioned maliciously are not discarded. Either these are bitstreams that are software files and are stored in the repository, or these can be physical hardware, which can only be discarded with manual intervention. Hence, physical proof is restored till forensic team analyses in post operation phases.

9.4.1.5 Direct Evidence

In general, direct evidence may be the username and IP address of the person doing illegal activities or sending objectionable files to or from a server.

For the present case, the sources that were associated previously with malicious operations can be considered cases for direct evidence. These are normally blacklisted temporarily, till further proof is found that actually confirms their malicious intent.

Questions

1. Explain cybercrime and justify how it can affect the security CIA triad.
2. Explain the levels of forensics present in the cloud to thing continuumm of the IoT era.
3. Explain how an adversary can target the server or system resources.
4. Explain how a system can be used as a tool for crime.
5. Explain the phase of **acquisition** in digital forensic investigation, with context to reconfigurable hardware platforms.
6. Explain the phase of **identification** in digital forensic investigation, with context to reconfigurable hardware platforms.
7. Explain the phase of **evaluation** in digital forensic investigation, with context to reconfigurable hardware platforms.
8. Explain the phase of **presentation** in digital forensic investigation, with context to reconfigurable hardware platforms.
9. Explain the reasons that motivate the search for evidence for digital forensics in reconfigurable hardware platforms.
10. Explain the various types of evidence for a forensic investigation.

References

Årnes A (2017) Digital forensics. Wiley
Holt T, Bossler A (2015) Cybercrime in progress: theory and prevention of technology-enabled offenses. Routledge
Hou J, Li Y, Yu J, Shi W (2019) A survey on digital forensics in internet of things. IEEE Int Things J 7(1):1–15

Chapter 10
Tutorial for Hands on Reconfigurable Hardware: System Design, Attacks, Security

Objectives of The Chapter

- Starting with hardware description language (HDL) and understanding the different styles of HDL Programming.
- Basics of VHDL and Verilog with simple programs.
- Programs for generating Test Bench Waveforms to analysis and understanding functional checking with Waveforms.
- HDL Programs to develop finite state machines and XOR based encoding.
- Hardware Trojan Attacks affecting the CIA security triad (Confidentiality, Integrity, Availability).
- Demonstrating Design Time Solution with Key based Obfuscation with HDL.
- Demonstrating Runtime Solutions with Majority Polling and Self Aware Agent based Security with HDL.

10.1 Introduction

For designing on reconfigurable hardware, we can consider a front end and a back end design structure.

The front end is generally associated with register-transfer-level (RTL) designing. Initially, we need to visualise the user specifications and then develop a blueprint of the system architecture. Hardware description language (HDL) is used to develop the codes for the system architecture. The basic two hardware programming is done via VHDL (Very High Speed Integrated Circuit Hardware Description Language) (Bhasker 1994) or Verilog (Palnitkar 2003). Functional verification is done for the HDL codes. This comprises of the front end design.

K. Guha et al., *Cybersecurity for Reconfigurable Hardware Based Critical Infrastructures*,
https://doi.org/10.1007/978-3-031-67591-1_10

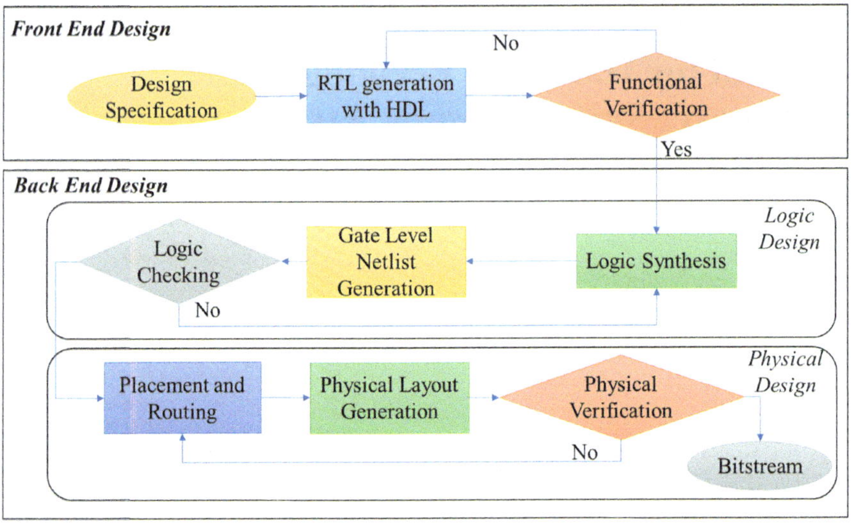

Fig. 10.1 Design flow

At the back end, we have logic designing and physical designing. For logic design, we have logic synthesis, followed by gate level netlist generation and logic equivalence check. This is generally carried forward to physical desing, where map, place and route operation takes place, followed by physical layout designing and verification to bitstream generation. This bitstream can be directly dumped into the reconfigurable hardware or FPGA fabric.

A diagrammatic representation of the entire procedure is shown in Fig. 10.1.

Different companies have different tools like Xilinx (ISE/ VIVADO), Alterra (Quartus) for the design of such architectures on reconfigurable hardware platforms. We need to write proper HDL codes, i.e. the RTL description and perform functional verification. Then the tool will generate the rest, i.e. gate level designing, synthesis, mapping, placement, routing and bitstream generation.

In this chapter, we will be essentially focusing on HDL generation, which is of key importance from the computer science point of view. We will also understand the points of vulnerability in such designs and how to overcome them.

10.2 Basic System Design

Essentially, system design with HDL is performed either with Verilog or VHDL (Very High Speed Integrated Circuit Hardware Description Language).

Moreover, there exists three style of programming, i.e. Dataflow, Behavioural and Structural, as depicted in Fig. 10.2.

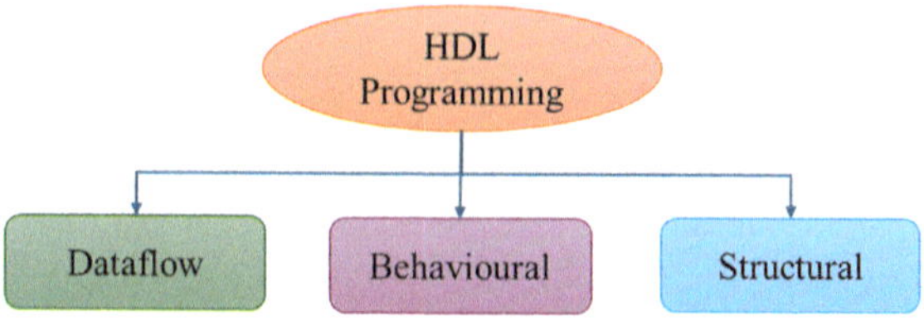

Fig. 10.2 HDL programming styles

Dataflow programming is associated with the flow of data from the input to output based on the various gates that it passes. Behavioural programming depends on if-then-else, case statements, do-while, etc. type of statements that essentially describes how a program behaves. Structural programming is associated with developing various components and then combining them to design the final design.

10.3 Getting Started with HDL Programming

HDL programs are associated with three key components. The first one is associated with defining of libraries and packages. The second part is associated with description of the various ports, i.e. inputs or outputs, which we also define as entity and finally is the architecture section, which is associated with description of the actual functional components.

We now describe both VHDL program and a Verilog program for a generic problem that can be solved with a simple basic gate. Then describe its truth table and how it is checked with the aid of a test bench waveform. For demonstration purposes, we use the EDA playground (https://edaplayground.com/home) that is available in the internet and is non-licensed.

10.3.1 Problem

Let's discuss a switch light problem. We have two switches and a bulb. We need to connect the switches and the bulb in such a manner that whenever one switch is enabled, the bulb will be on. The bulb will be off only when both the switches are disabled.

For this, we need to connect the two switches and the bulb with an OR gate. The two switches are the two inputs to the OR gate, while the bulb is the output. The circuit diagram is depicted in Fig. 10.3.

The possible truth table for this type of problem is depicted in Table 10.1.

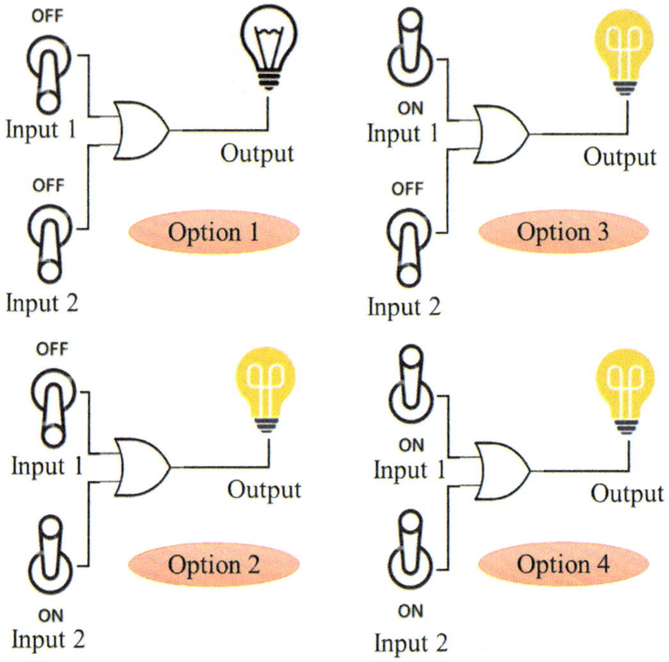

Fig. 10.3 Light switch problem

Table 10.1 Operations for the problem or basic OR gate

Input 1 (a)	Input 2 (b)	Output (q)
0	0	0
0	1	1
1	0	1
1	1	1

10.3.2 Writing a VHDL Program

10.3.2.1 Overview

VHDL programs has an explicit way of dividing the program into three parts:

(i) Defining Libraries and Packages: Like normal programming, it is not possible to program in HDL from scratch. Hence, we need to import and use libraries and packages that aid us and simplifies our hardware programming experience.

(ii) Defining the Entity: An entity can be termed as a specification of the design's external connections or interfaces. Entity declarations notifies the following:

1. Name of the entity
2. Set of port declarations defining the inputs and outputs to the hardware design

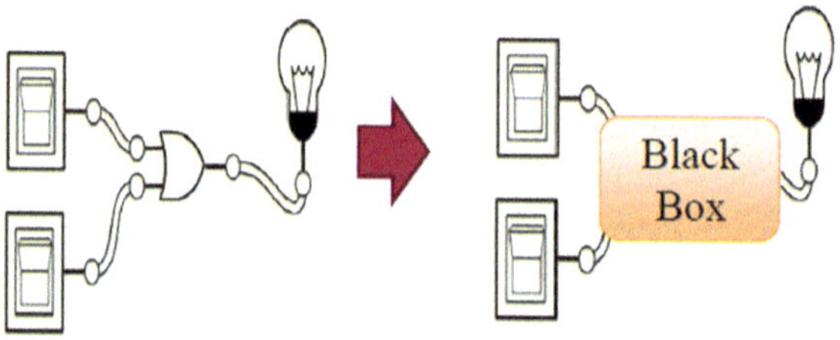

Fig. 10.4 Depicting black box concept for architecture of the discussed switch light problem

(iii) Defining the architecture: An architecture is a specification of the design's internal implementation or what is present inside the black box, as depicted in Fig. 10.4.

10.3.2.2 Main Program

VHDL program for an *OR* gate operation is shown in Fig. 10.5. In this program, the name of the entity is *or_gate*, where the input ports are *a* and *b* and the output port is *q*. The name of the architecture is *rtl*. <= denotes the assignment operation and *or* keyword defines the *OR operation*.

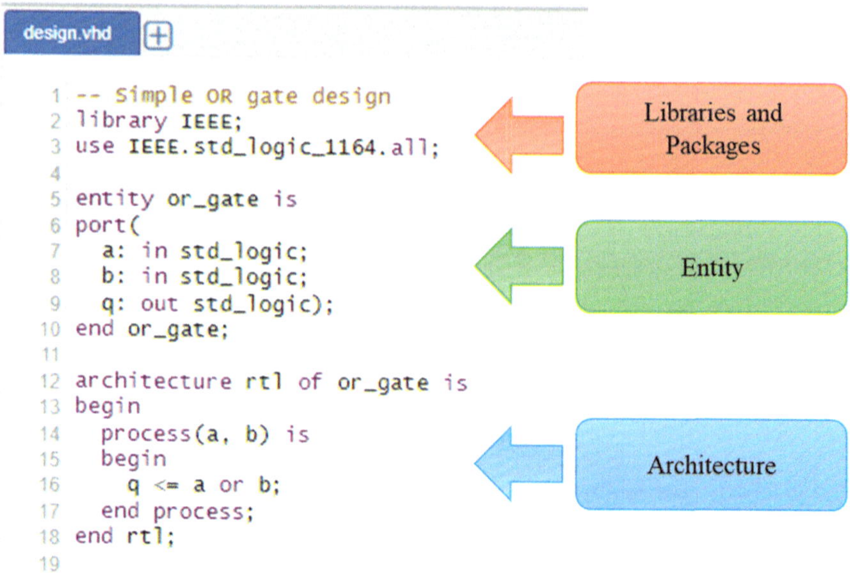


```
-- Simple OR gate design
library IEEE;
use IEEE.std_logic_1164.all;

entity or_gate is
port(
  a: in std_logic;
  b: in std_logic;
  q: out std_logic);
end or_gate;

architecture rtl of or_gate is
begin
  process(a, b) is
  begin
    q <= a or b;
  end process;
end rtl;
```

Fig. 10.5 VHDL program for OR gate operation

10.3.2.3 Program for Generating Test Bench Waveform

After writing an HDL program, we need to check its functional correctness. For this, a test bench waveform is required. A test bench waveform for the OR gate operation that is written in VHDL is depicted in Fig. 10.6.

testbench.vhd

```
-- Testbench for OR gate
library IEEE;
use IEEE.std_logic_1164.all;

entity testbench is
-- empty
end testbench;

architecture tb of testbench is

-- DUT component
component or_gate is
port(
  a: in std_logic;
  b: in std_logic;
  q: out std_logic);
end component;

signal a_in, b_in, q_out: std_logic;

begin

  -- Connect DUT
  DUT: or_gate port map(a_in, b_in, q_out);

  process
  begin
    a_in <= '0';
    b_in <= '0';
    wait for 1 ns;
    assert(q_out='0') report "Fail 0/0" severity error;

    a_in <= '0';
    b_in <= '1';
    wait for 1 ns;
    assert(q_out='1') report "Fail 0/1" severity error;

    a_in <= '1';
    b_in <= 'X';
    wait for 1 ns;
    assert(q_out='1') report "Fail 1/X" severity error;

    a_in <= '1';
    b_in <= '1';
    wait for 1 ns;
    assert(q_out='1') report "Fail 1/1" severity error;

    -- Clear inputs
    a_in <= '0';
    b_in <= '0';

    assert false report "Test done." severity note;
    wait;
  end process;
end tb;
```

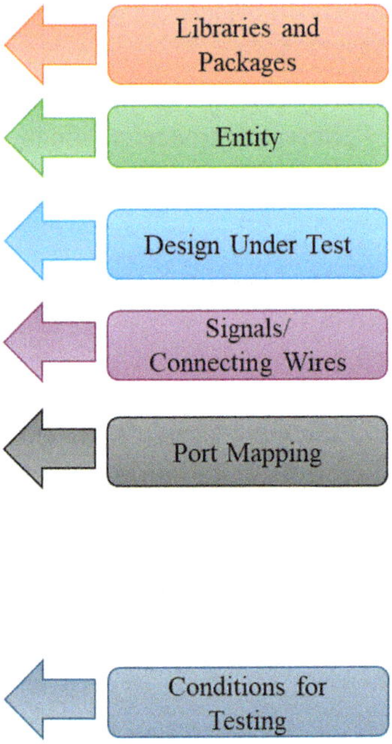

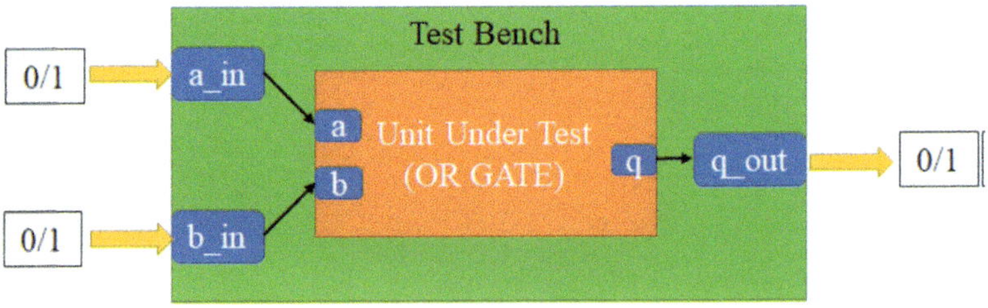

Fig. 10.6 Test bench waveform with VHDL for OR gate program

Like a main VHDL program, a test bench waveform for a VHDL program also starts with importing of libraries and packages.

However, the entity for a test bench waveform is empty. This is because a testbench is a closed system and all inputs for a particular design under test (DUT) are driven by the testbench. Even all the outputs for the DUT is observed by the testbench. So, the testbench is devoid of any inputs or outputs. The name of the entity for this particular case is *testbench*.

After this, the architecture starts, which is named *tb* for the present scenario. The architecture starts with importing the design to be tested as a component over here. Presently, the imported design is *or_gate*. If overserved carefully, the entity of the *or_gate* is simply re-written over here, which contains all the ports.

After this, the signals are defined. Signals in VHDL are connecting wires. Over here, the ports of the DUT needs to be connected with the ports of the test bench. This is formally done in the port mapping.

Post port mapping, all the various conditions for testing are provided, along with the timing constraints.

10.3.3 Writing a Verilog Program

Let us view how the same Switch Light Problem or the OR gate program is written in Verilog.

10.3.3.1 Main Program

In Verilog, there is no explicit dividing of a program like VHDL. However, like VHDL we define the inputs and outputs or the ports at the begining. This is followed by description of the working or architecture. Figure 10.7 depicts a Verilog program for the switch light problem.

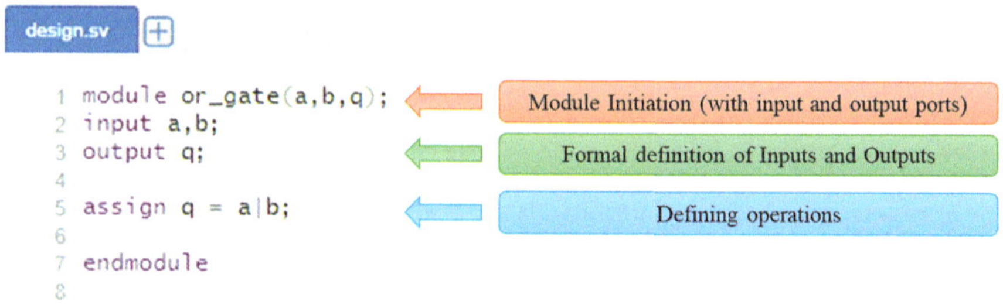

Fig. 10.7 Verilog program for the OR gate operation

10.3.3.2 Program for Generating Test Bench Waveform

Like VHDL, we can also write a program to generate Test Bench Waveform with Verilog. Figure 10.8 illustrates this. The program has a different module name, as it is a test bench generating Verilog program and its name should not be same as that of the original program. The ports or inputs/outpus of the test bench are defined initially, followed by incorporating the actual program module and connecting the related ports, i.e. port mapping. Then, the testing confitions with various timing constraints are defined. It is to be noted that # does the same work in Verilog as command *wait for* does in VHDL.

10.3.4 Checking Functional Correctness with Test Bench Waveform

For the present switch light problem, which is essentially an OR gate operation, both the VHDL and Verilog will produce the same results.

The test bench waveform is obtained in the EP Wave in the EDA Playground, as depicted in Fig. 10.9. We illustrate it for the VHDL ports. As the main VHDL program ports and the test bench ports are directly connected to each other and there are no other functionalities in the program, hence, they are identical. As evident from the program, the output is 0, when both inputs are 0 and if any one of the inputs is 1, the output is 1.

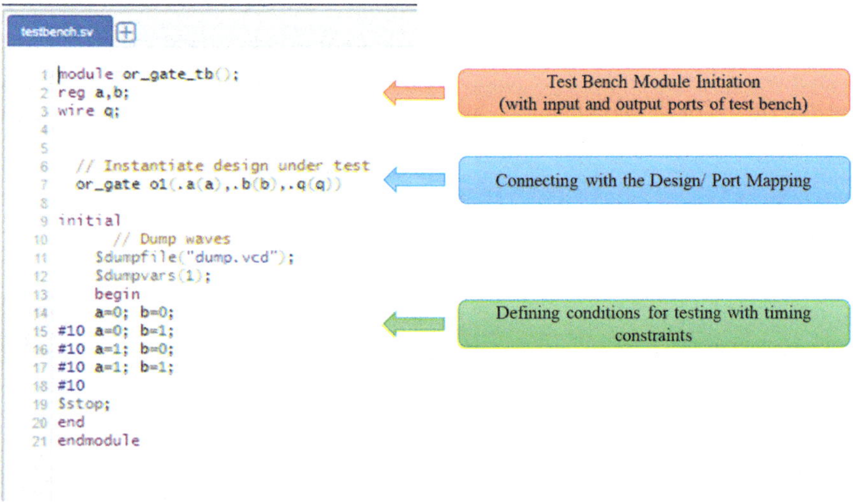

Fig. 10.8 Test bench waveform with verilog for OR gate program

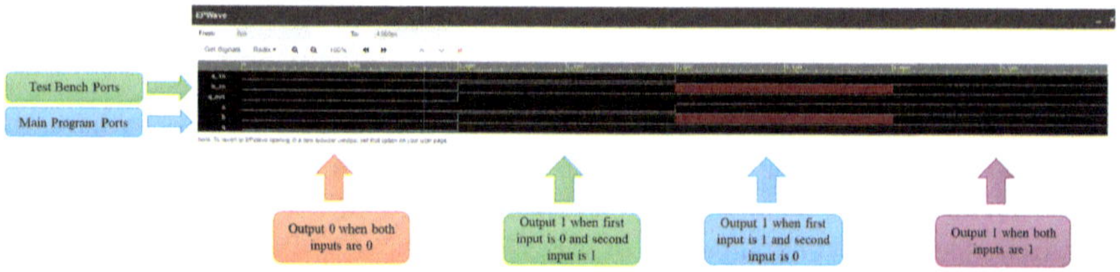

Fig. 10.9 Test bench waveform (obtained in EP Wave)

10.4 Different Forms of HDL Programming

In this section, we will illustrate the different forms of HDL programming, as discussed previously. For this, we use a 2:1 multiplexer, whose circuit diagram is provided in Fig. 10.10. This comprises of two basic inputs (A and B) and a select line (S). A, B, S are three input ports and there is an output port (Y). The value of Y depends on the value of S. If S = 0, then Y will reflect the value of A, else if S = 1, Y will relect the value of B.

10.4.1 Dataflow

Dataflow programming essentially represents the flow of data from input ports to the output port. This comprises either a single or a group of statements representing basic digital gate operations. A dataflow description in VHDL for the 2:1 Mux is shown in Fig. 10.11.

Fig. 10.10 Circuit diagram for 2:1 Mux

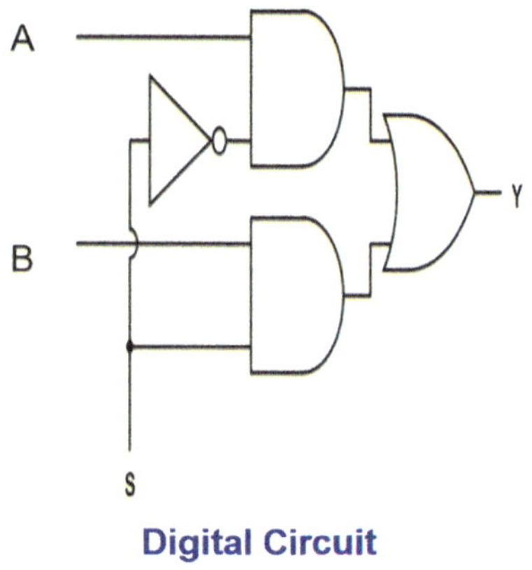

10.4.2 Behavioural

Behavioural programming is associated with describing a process, where the behaviour of the operations are depicted. It generally comprises of statements like *if*, *then*, *else*, *switch* statements, etc. This also comprises of a sensitivity list. The sensitivity list (which contains a set of variables) of a particular process implies that the operations present within the process will only be executed if any of the variables in the sensitivity list is associated with a change. For the VHDL based behavioural program of 2:1 mux, depicted in Fig. 10.12, comprises of a sensitivity list of 3 input variables, a, b and s. If any of these are associated with a change in value, then the *if*, *then*, *else* statements within the process will be executed that will lead to a change in the value of output y.

10.4.3 Structural

This style of programming is associated with building a big program with small components or sub-programs. Essentially, the components are duplicated and integrated to develop the entire program. For the 2:1 mux, three components are defined, i.e. *and*2, *not*1, *or*2. These are integrated in the architecture block, via port mapping the components for depicting different types of operations. This is depicted in Fig. 10.13.

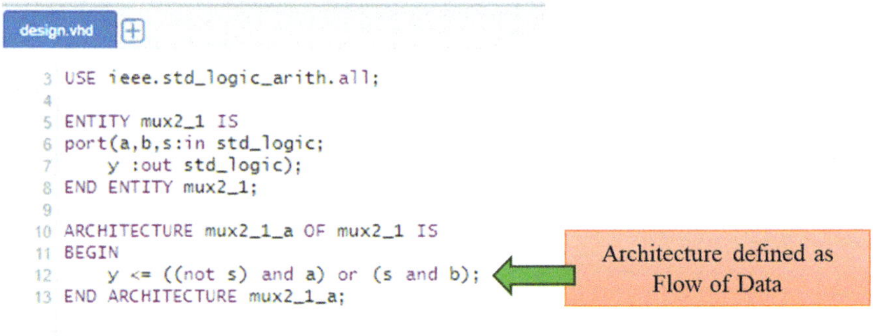

Fig. 10.11 Dataflow programming for 2:1 Mux

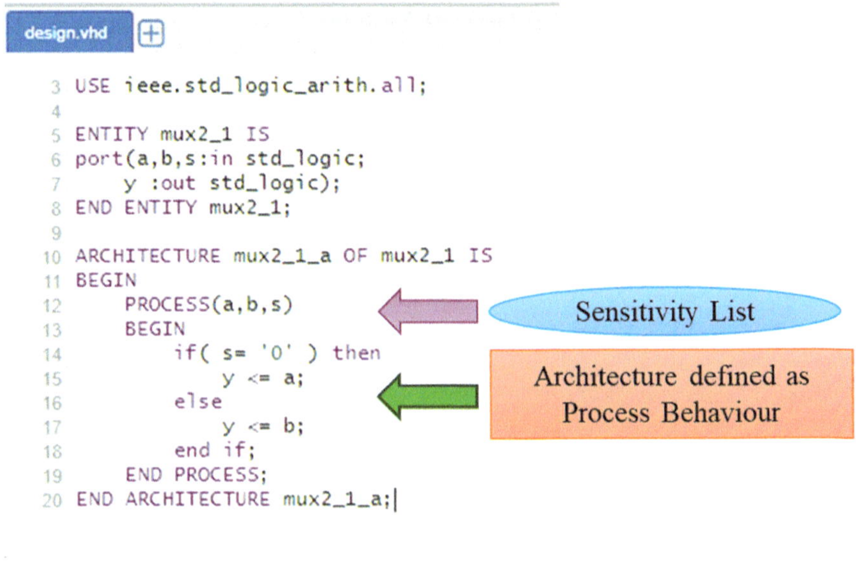

Fig. 10.12 Behavioural programming for 2:1 Mux

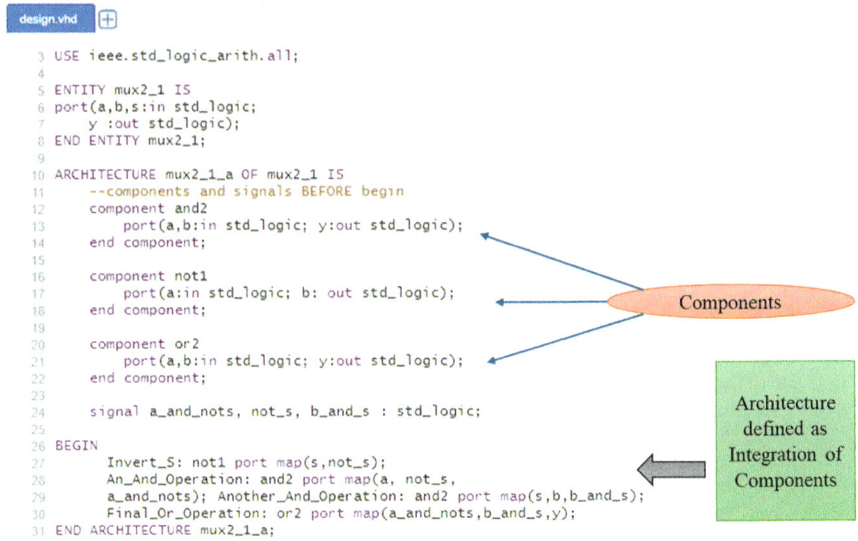

Fig. 10.13 Structural programming for 2:1 Mux

10.4.4 Checking Functional Correctness

This is associated with developing the program for generating the test bench waveform and analyzing the test bench waveform. The program for the test bench waveform is depicted in Fig. 10.14 and the waveform is shown in Fig. 10.15. As evident from the waveform, when the value of s is 0, the value of y reflects the value of a and when the value of s is 1, the value of y reflects the value of b.

testbench.vhd

```
-- Testbench for 2:1 mux gate
library IEEE;
use IEEE.std_logic_1164.all;
entity testbench is
-- empty
end testbench;
architecture tb of testbench is
-- UUT component
component mux2_1 is
port(
a,b,s:in std_logic;
y :out std_logic);
end component;
signal a_in, b_in, s_in, y_out: std_logic;
begin
-- Connect UUT
DUT: mux2_1 port map(a_in, b_in, s_in, y_out);
process
begin
a_in <= '1';
b_in <= '0';
s_in <= '0';
wait for 10 ns;
a_in <= '1';
b_in <= '0';
s_in <= '1';
wait for 10 ns;
-- Clear inputs
a_in <= '1';
b_in <= '0';
s_in <= '0';
wait for 10 ns;
wait;
end process;
end tb;
```

Fig. 10.14 Program for generating a test bench waveform of 2:1 Mux

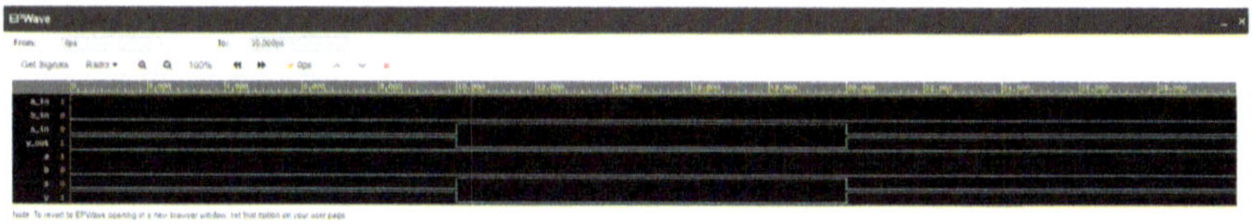

Fig. 10.15 Test bench waveform for 2:1 Mux

10.5 System Designs with Verilog

10.5.1 Designing an FSM Accumulator with Verilog

In this subsection,we will depict how Verilog can be used to design a finite state machine (FSM) that basically accumulates or sums the inputs coming in each of its state. As depicted in Fig. 10.16, the FSM comprises a total of five operational states, starting from $S1$ and moves on to $S5$, transiting in each clock cycle. Initially, the accumulator value (acc) is null. At each state, the input values are summed with the existing acc. After $S5$ is reached, the output (out) provides the total accumulated value.

The Verilog program is shown in Fig. 10.17. The input comprises of a clock (clk) and reset ($reset$) signals, along with the five inputs ($input1$, $input2$, $input3$, $input4$, $input5$) for the five states. There are two outputs, the accumulator value at each and every state can be found from $acc_{o}ut$, while the end result can be seen from out. There are three intermediate registers acc, acc_x, $state$ to store intermediate values. We consider operations with 4 bit values throughout.

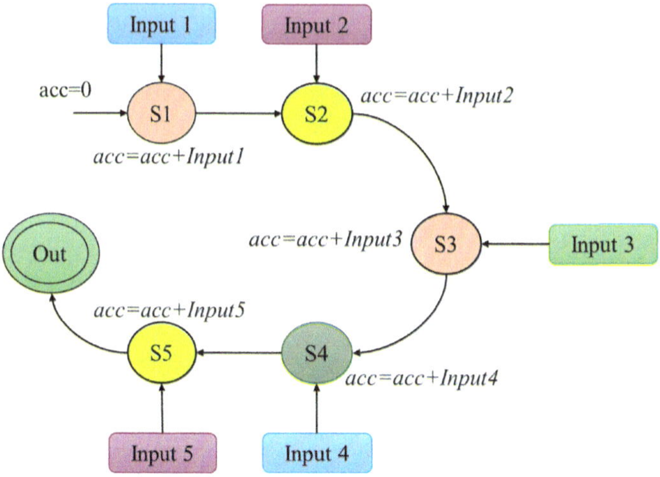

Fig. 10.16 Diagrammatic representation of FSM

design.sv

```
module fsm_accumulate(input clk, reset,
                          [3:0] input1, [3:0] input2, [3:0]
                          input3, [3:0] input4, [3:0] input5,
                          output[3:0] acc_out, [3:0] out);

  reg [3:0] acc;
  reg [3:0] acc_x;
  reg [3:0] state;


always @(posedge clk or posedge reset)
begin
if(reset)
    begin
      acc <= 4'd0;
      acc_x <= 4'd0;
      state <= 4'd1;
    end

  if(state ==1 )
    begin
      acc <= acc + input1;
        state <= state +4'd1;
    end

  if(state ==2 )
    begin
      acc <= acc + input2;
        state <= state +4'd1;
    end
  if(state ==3 )
    begin
      acc <= acc + input3;
        state <= state +4'd1;
    end

  if(state ==4 )
    begin
      acc <= acc + input4;
        state <= state +4'd1;
    end

  if(state ==5 )
    begin
      acc <= acc + input5;
        state <= state +4'd1;
    end

  if (state>5)
          acc_x = acc;
end
  assign acc_out=acc;
  assign out=acc_x;
endmodule
```

Fig. 10.17 Verilog program for FSM

testbench.sv

SV/Verilog Testbench

```
module fsm_accumulate_testbench();
    reg clk, reset;
    wire [3:0] acc_out;
    wire [3:0] out;
    reg [3:0] input1, input2, input3, input4, input5;

    fsm_accumulate dut(clk, reset, input1, input2, input3, input4, input5, acc_out,out);

    initial begin
    // Dump waves
    $dumpfile("dump.vcd");
    $dumpvars(1);
    input1=4'd1; input2=4'd2; input3=4'd3; input4=4'd4; input5=4'd5;
    clk=0;
    always #5 clk=~clk;

    end
    initial begin
    reset=1;
    #10;
    reset=0;
    #70;
    reset=1;
    #10;
    reset=0;
    #70;
    reset=1;
    end
endmodule
```

Fig. 10.18 Test bench program in verilog for FSM

When *reset* is set or 1, all the intermediate registers are set to 0. Else, with each clock cycle, the FSM proceeds one state at a time. Operations in each state is almost the same. The value of *acc* is summed with the respective input at that state. When the final state is reached, the output is generated. The values of the accumulator at each state is reflected by *acc_out* and the final output is reflected by *out*.

For generating the test bench waveform also, a Verilog program is used. A sample is depicted in Fig. 10.18. As discussed before, the test bench module is empty, i.e. does not comprise of any inputs or outputs. It takes the main program, i.e. *fsm_accumulate* as the design under test (*dut*). Related registers are defined. The five inputs are defined and they are static or does not change with time. However, the clock repeats after every 5 time units and the reset is set at fixed time intervals, as shown in the figure.

The waveform is shown in Fig. 10.19. The increment in the values of the accumulator can be seen from *acc_out*, in each state. While the value of *out* depicts the final output that is reached after the five states. It can also be seen that with *reset* signal, all the intermediate registers and the value of the accumulator is set to 0.

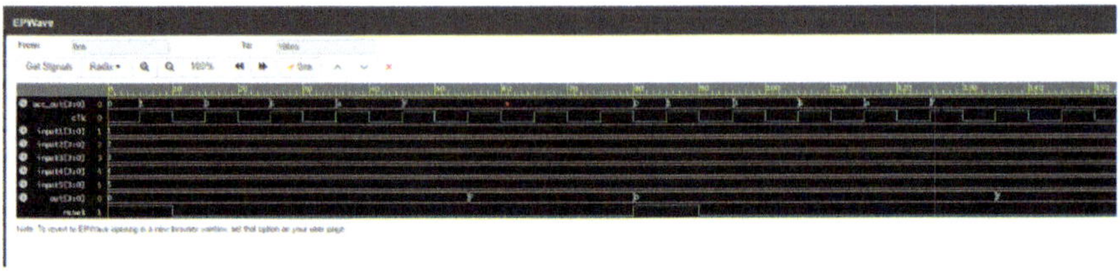

Fig. 10.19 Test bench waveform for FSM

10.5.2 Encoding FSM Accumulator's Output

In this subsection, we demonstrate encoding of the result of the FSM discussed in the previous subsection, based on a secret key.

Encoding-decoding can be carried out via different innovative algorithms. However, for simplicity, we demonstrate how XOR based encryption-decryption can be carried out.

As per XOR logic, if both operands are same, then the output is 0, else if the operands are different, the output is 1. This is depicted in Fig. 10.20. As illustrated, in the encryption side, if the plaintext is 1101 and the secret key is 0101, then the ciphertext or output is 1000. Similarly, in decryption side, when ciphertext is 1000 and XOR logic operation is performed with secret key is 0101, then we get back the plaintext, i.e. 1101.

We consider the FSM depicted in Fig. 10.16. However, we add a new state there, S6. In S6, the XOR operation is performed with the secret key. After the operation in S6, the output is given out. This is shown in Fig. 10.21.

Figure 10.22 depicts the verilog program of the FSM, with encoding with XOR operation. This is quite similar to the previous one, except that a secret key is present. This is assigned to a register. In state S6, the secret key is XOR ed with the value of the accumulator.

The test bench program is same as the previous one, i.e. Fig. 10.18. Hence, we do not depict it again in this subsection. The test bench waveform is shown in Fig. 10.23.

Fig. 10.20 XOR encryption-decryption logic

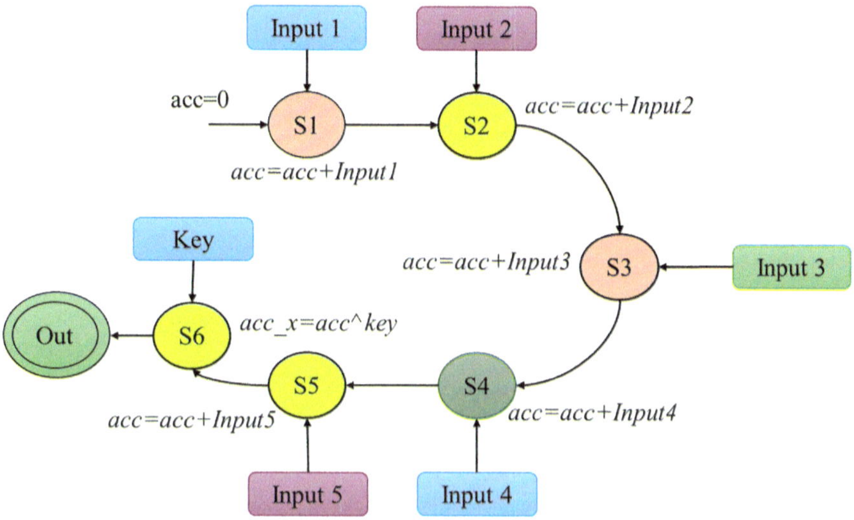

Fig. 10.21 FSM diagram with XOR encoding

As illustrated in Fig. 10.23, the output *out*, does not reflect the final value of the accumulator, but the XOR ed value of the secret key and the final value of the accumulator.

10.6 Attacks

Attacks can be created in many ways, as described in previous chapters. For demonstration in this chapter, we will be showing the effect of hardware trojan horses (HTHs) on the designed HDL programs in the previous sections of this chapter.

A HTH is a malicious circuit element that is inserted in the hardware design by a malicious designer. Moreover, when different third party intellectual properties (3PIPs) are procured and integrated, then scope lies in some of the 3PIPs being malicious.

The structure of a HTH essentially comprises a trigger and a payload. HTHs may be always on or get activated at certain conditions. The trigger is optional for HTHs that are always on. However, the achitecture of the trigger determines the nature of HTHs that get activated at a certain point of time.

The trigger may be internal or external. Internal trigger may be a time bomb (counter till a desired value is reached) or certain rare combination of internal node values. External trigger comprise signals captured from a remote source via sensors or antenna. The structure of trigger essentially depends on the adversary who designs it.

design.sv

```
module fsm_accumulate(input clk, reset,
                      [3:0] input1, [3:0] input2, [3:0]
                       input3, [3:0] input4, [3:0] input5,
                       output[3:0] acc_out, [3:0] out);
  reg [3:0] acc;
  reg [3:0] acc_x;
  reg [3:0] state;

  reg [3:0] key=4'b0101;

always @(posedge clk or posedge reset)
begin
if(reset)
    begin
      acc <= 4'd0;
      acc_x <= 4'd0;
      state <= 4'd1;
    end

  if(state ==1 )
    begin
      acc <= acc + input1;
        state <= state +4'd1;
    end

  if(state ==2 )
    begin
      acc <= acc + input2;
        state <= state +4'd1;
    end
  if(state ==3 )
    begin
      acc <= acc + input3;
        state <= state +4'd1;
    end

  if(state ==4 )
    begin
      acc <= acc + input4;
        state <= state +4'd1;
    end

  if(state ==5 )
    begin
      acc <= acc + input5;
        state <= state +4'd1;
    end

  if(state ==6 )
    begin
      acc_x <= acc ^ key;
        state <= state +4'd1;
    end
 end
  assign acc_out=acc;
  assign out=acc_x;
endmodule
```

Fig. 10.22 Verilog program for FSM with XOR encoding

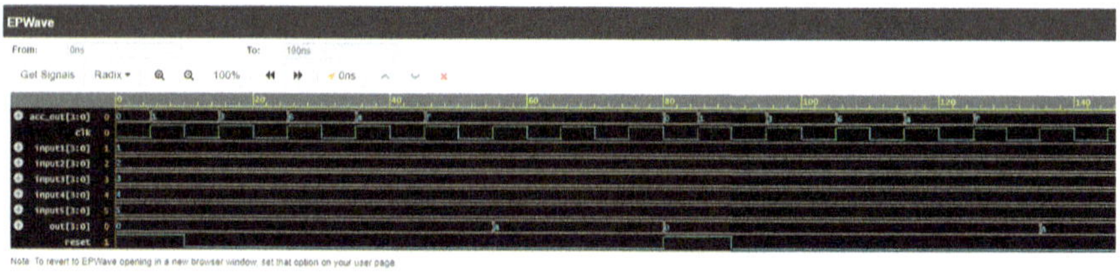

Fig. 10.23 Test bench waveform for FSM with XOR encoding

The payload comprises of the malicious functionality. This may affect the basic security primitives of the system, i.e. CIA (confidentiality, integrity, availability) properties.

10.6.1 Affect Integrity

We consider the 2:1 mux (depicted in Fig. 10.10), for demonstrating the effect of integrity attack by a HTH. Let the HTH be an always on HTH and we neglect the trigger. The payload can be any simple additional logic gate that will change the functionality. Such a malicious circuit element can be present in any place of the digital circuit. For ease of demonstration and simplicity of understanding, we consider a simple AND gate with logic 0 that is placed just before output. This is depicted in Fig. 10.24.

The HDL code that is written in Verilog, which comprises of HTH is depicted in Fig. 10.25. The intermittent signal $y1$ is passed through logic AND operation with logic 0. The test bench waveform is same as that of Fig. 10.14. The final output is 0 throughout and does not vary as shown in Fig. 10.26.

Fig. 10.24 Integrity attack by HTH on 2:1 Mux

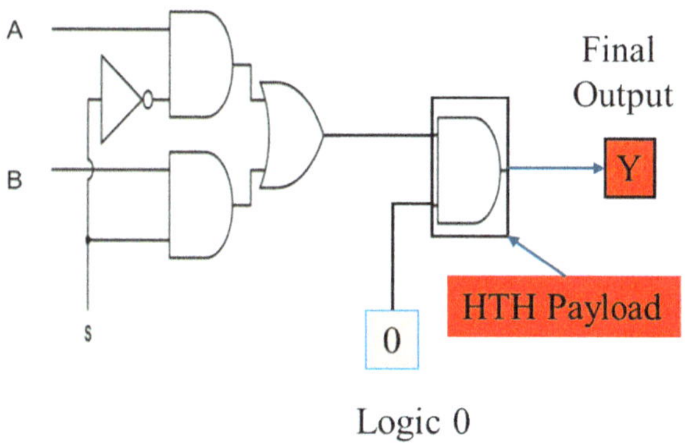

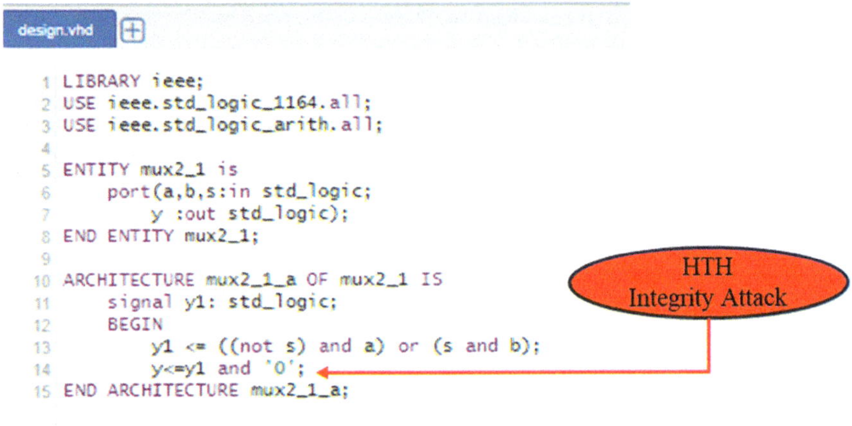

Fig. 10.25 Verilog code for integrity attack by HTH

Fig. 10.26 Waveform demonstrating integrity attack

10.6.2 *Affect Confidentiality*

We consider the design of Fig. 10.21 to demonstrate confidentiality attack by a HTH. Working of the HTH takes place after the XOR encryption in state S6, as shown in Fig. 10.27. The confidentiality attack is associated with revealing the secret key. This is done via analysis of the side channel parameter, timing or delay.

If the key bit is 0, no delay is associated in getting the output. However, if the key bit is 1, then an additional delay of 2 time units is associated in generating the result. This is illustrated in Fig. 10.27. The delay can be caused via some buffers or combination of two NOT gates or via using some dummy circuitry. Related verilog code is depicted in Fig. 10.28. Associated waveform is shown in Fig. 10.29. As evident, when the key bit is 0, no delay in output is seen. However, when the key bit is 1, a delay of 2 time units is seen.

10.6.3 *Affect Availability*

We consider the FSM described in Fig. 10.16 for demonstrating availability attack. The HTH considered is always on and hence, no trigger is represented. The HTH is

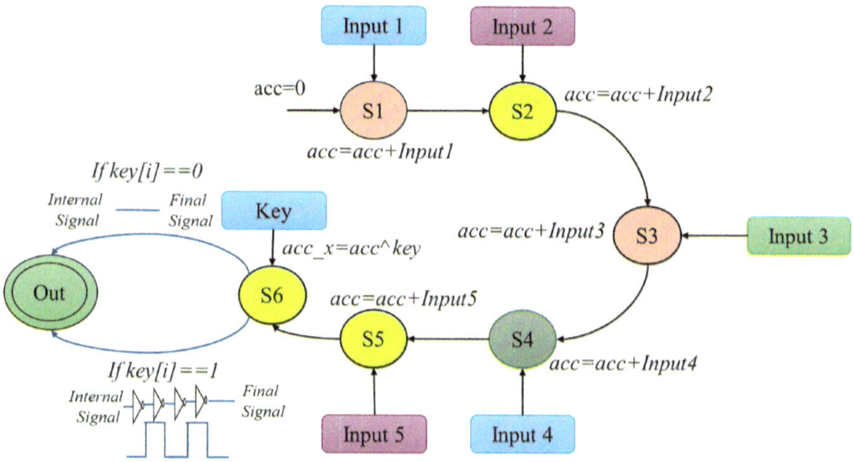

Fig. 10.27 Confidentiality attack by HTH on FSM with encoding

placed just before the output. Thus, after state S5, the HTH works on the output of S5. Delays can be of two types. One is the finite delay and another infinite delay. Finite delays are caused by buffers, or combination of two NOT gates, or some dummy operations that operates for some fixed clock cycles. However, infinite delays are caused by some delay causing logic that repeats every clock cycle. The basic diagram is depicted in Fig. 10.30.

Corresponding verilog code is depicted in Fig. 10.31. Related waveforms is shown in Fig. 10.32. Let deadline be at 54 time units. The finite delay causes output generation at 74 time units. However, for the infinite delay, no output is produced. Thus, availability attack is caused.

10.7 Solution

10.7.1 Design Time Solution: Key Based Obfuscation

Design time solutions are associated with embedding of secret parameters in the design, so that adversaries in the foundry cannot understand. One such is embedding of keys. In this, the logic is locked by the designer, so that adversaries in the foundry may not get access to the functionalities, via various side channels. The keys are secretly shared to valid customers. At runtime, valid customers may deploy the keys and get access to the functionalities.

For demonstration, we consider the FSM demonstrated in Fig. 10.16. Each state is associated with a functionality. If key based obfuscation is used for this type of design, the functionality in each state can only take place if the right key is applied.

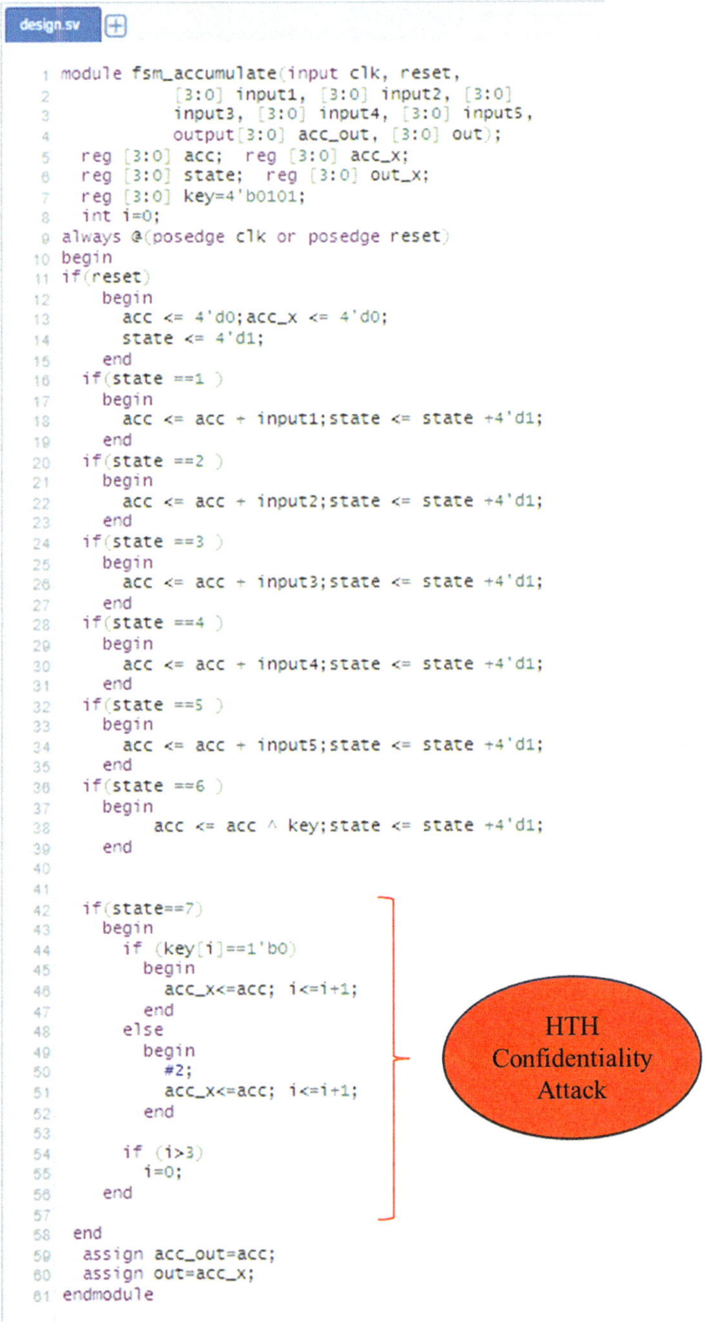

```
design.sv
module fsm_accumulate(input clk, reset,
              [3:0] input1, [3:0] input2, [3:0]
              input3, [3:0] input4, [3:0] input5,
              output[3:0] acc_out, [3:0] out);
  reg [3:0] acc;  reg [3:0] acc_x;
  reg [3:0] state;  reg [3:0] out_x;
  reg [3:0] key=4'b0101;
  int i=0;
always @(posedge clk or posedge reset)
begin
if(reset)
      begin
        acc <= 4'd0;acc_x <= 4'd0;
        state <= 4'd1;
      end
  if(state ==1 )
      begin
        acc <= acc + input1;state <= state +4'd1;
      end
  if(state ==2 )
      begin
        acc <= acc + input2;state <= state +4'd1;
      end
  if(state ==3 )
      begin
        acc <= acc + input3;state <= state +4'd1;
      end
  if(state ==4 )
      begin
        acc <= acc + input4;state <= state +4'd1;
      end
  if(state ==5 )
      begin
        acc <= acc + input5;state <= state +4'd1;
      end
  if(state ==6 )
      begin
            acc <= acc ^ key;state <= state +4'd1;
      end


  if(state==7)
      begin
        if (key[i]==1'b0)
          begin
            acc_x<=acc; i<=i+1;
          end
        else
          begin
            #2;
            acc_x<=acc; i<=i+1;
          end

        if (i>3)
          i=0;
      end

 end
  assign acc_out=acc;
  assign out=acc_x;
endmodule
```


Fig. 10.28 Verilog code demonstrating confidentiality attack

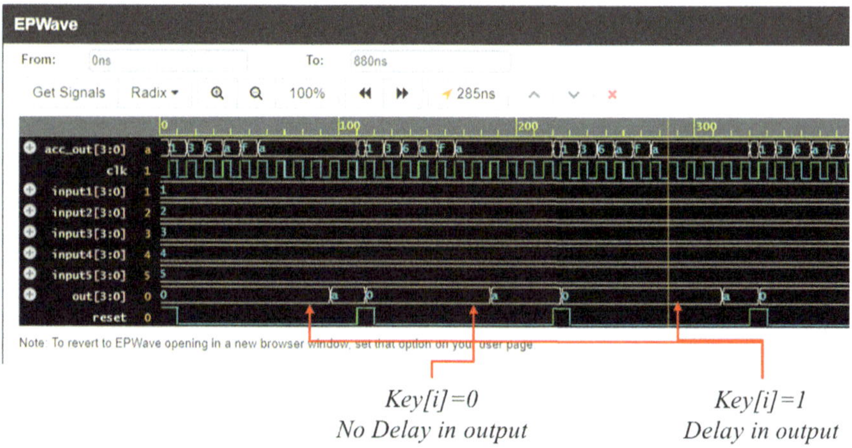

Fig. 10.29 Waveform demonstrating confidentiality attack

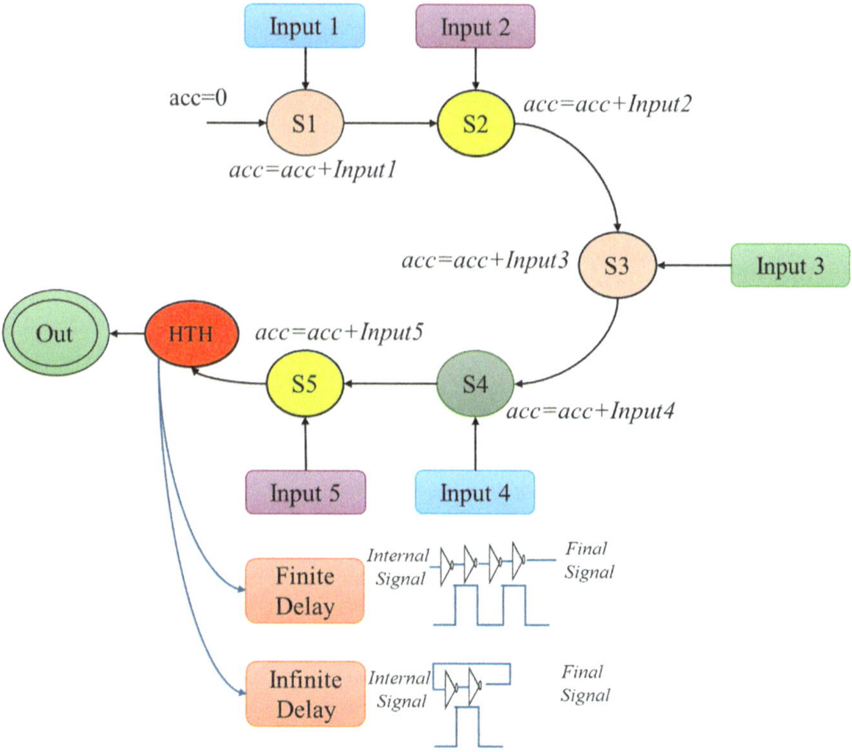

Fig. 10.30 Availability attack by HTH on FSM

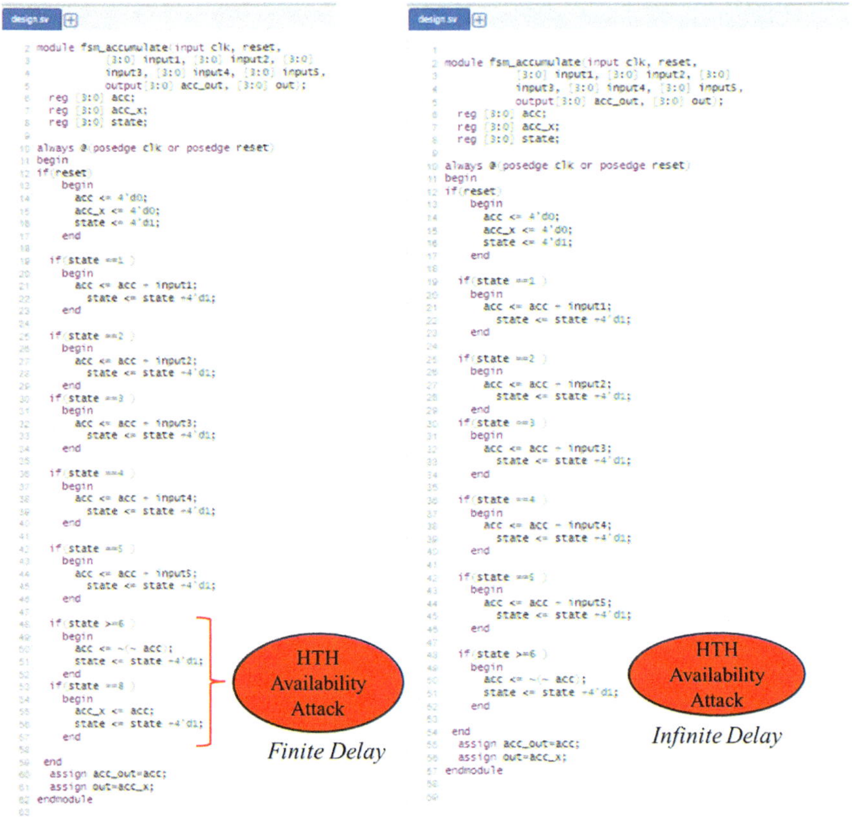

Fig. 10.31 Verilog code demonstrating availability attack

Figure 10.33 depicts the FSM Accumulate design, along with keys. The complexity can be increased based on the nature of keys. The verilog code is shown in Fig. 10.34. In each state, the corresponding key is checked. If the key is fine, then working of the state takes place, else the system stops working. This can be seen from the waveforms depicted in Fig. 10.35.

10.7.2 Runtime Solution

Runtime solutions are generally applied in the post-deployment phases. They are also termed as the last-line-of-defence. Mostly, these are associated with external monitoring modules or agents that observes or monitors the operations and takes a decision at runtime.

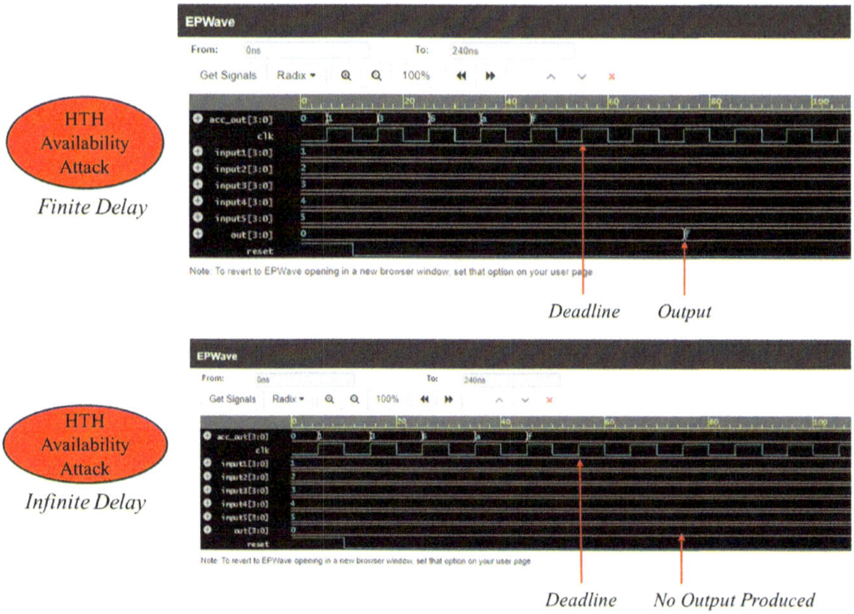

Fig. 10.32 Waveform demonstrating availability attack

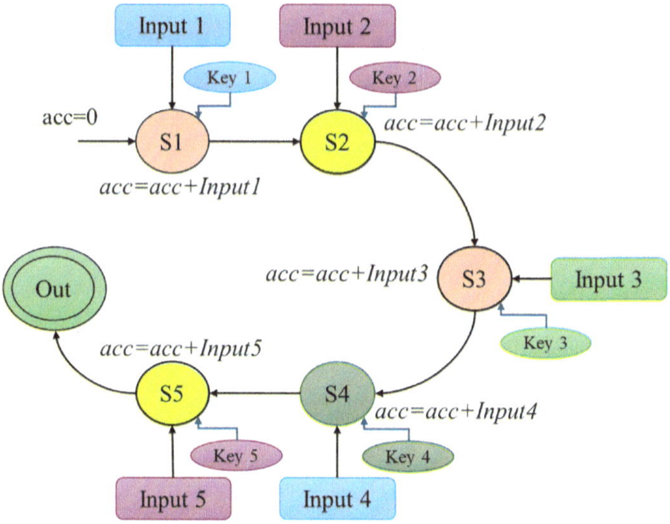

Fig. 10.33 FSM accumulate with keys

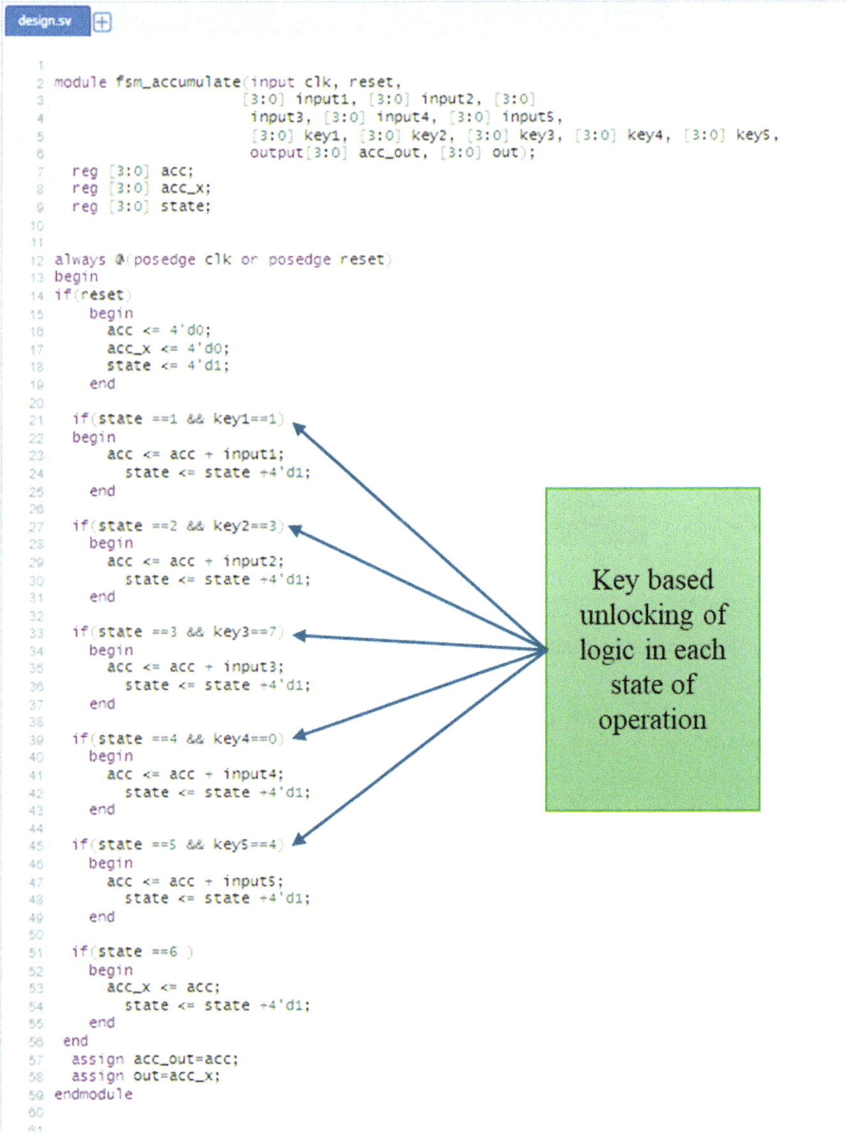

```
module fsm_accumulate(input clk, reset,
                      [3:0] input1, [3:0] input2, [3:0]
                      input3, [3:0] input4, [3:0] input5,
                      [3:0] key1, [3:0] key2, [3:0] key3, [3:0] key4, [3:0] key5,
                      output[3:0] acc_out, [3:0] out);
  reg [3:0] acc;
  reg [3:0] acc_x;
  reg [3:0] state;


always @(posedge clk or posedge reset)
begin
if(reset)
    begin
      acc <= 4'd0;
      acc_x <= 4'd0;
      state <= 4'd1;
    end

  if(state ==1 && key1==1)
  begin
      acc <= acc + input1;
        state <= state +4'd1;
    end

  if(state ==2 && key2==3)
    begin
      acc <= acc + input2;
        state <= state +4'd1;
    end

  if(state ==3 && key3==7)
    begin
      acc <= acc + input3;
        state <= state +4'd1;
    end

  if(state ==4 && key4==0)
    begin
      acc <= acc + input4;
        state <= state +4'd1;
    end

  if(state ==5 && key5==4)
    begin
      acc <= acc + input5;
        state <= state +4'd1;
    end

  if(state ==6 )
    begin
      acc_x <= acc;
        state <= state +4'd1;
    end
 end
  assign acc_out=acc;
  assign out=acc_x;
endmodule
```


Fig. 10.34 Verilog code to demonstrate FSM accumulate with keys

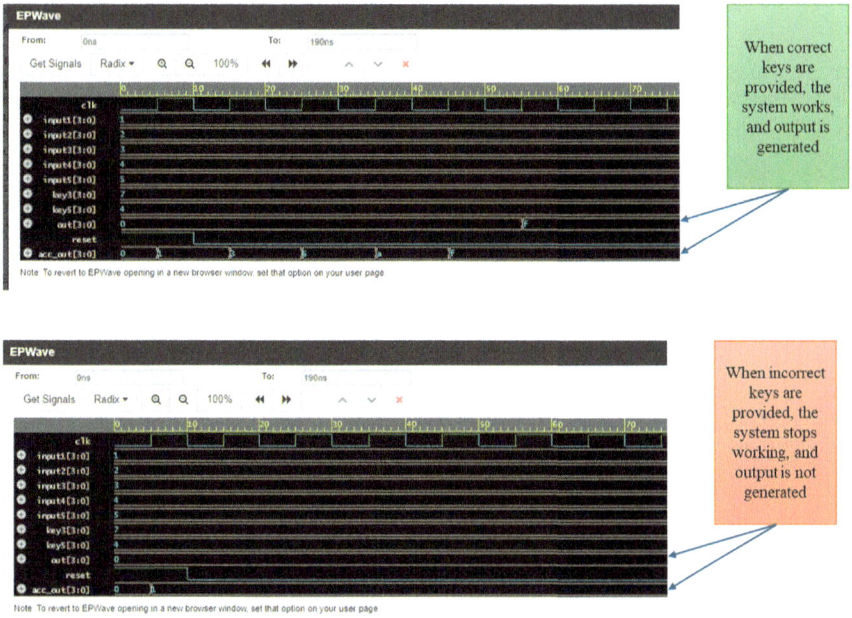

Fig. 10.35 Waveforms to demonstrate FSM accumulate with keys

Such types of defences can be broadly categorized as (i) redundancy based solutions (ii) self aware solutions.

10.7.2.1 Redundancy Based Solution

Redundancy based solutions work on the fact that the same type of malware will not be present in intellectual properties that are procured from different sources. A diagrammatic representation of the redundancy based solution is depicted in Fig. 10.36.

For instance, lets consider the integrity attack on a 2:1 mux. In such a case, if three intellectual properties (IPs) that performs the same operation, i.e. operation of the 2:1 mux is procured from three different vendors (like vendor 1 supplies the 2:1 mux IP in dataflow type, vendor 2 supplies the 2:1 mux IP in behavioural type and vendor 3 supplies the 2:1 mux IP in structural type), then the probability that the same malware will be present in all the three intellectual properties is bare minimum or negligible. Figure 10.37 illustrates this. Even if their payload or the type of malfunction is the same, then at least their time of trigger will be different. Finally, the result is chosen via a majority polling of the results. To sum up, it can be concluded that the same malfunction will not be applicable from resources at the same time that are procured from diverse sources.

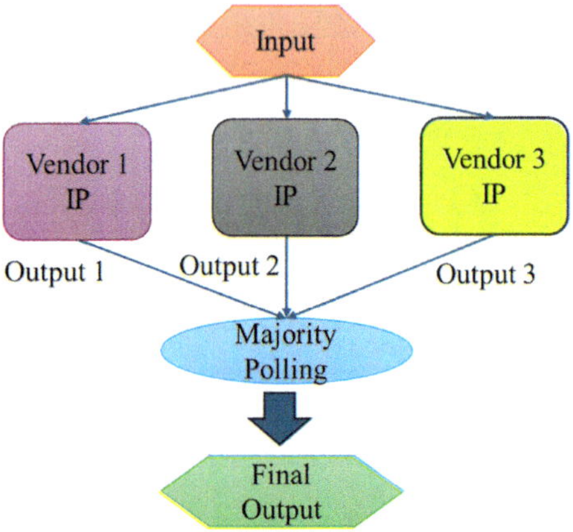

Fig. 10.36 Redundancy based solution

design.vhd

```
library IEEE;
Use IEEE. STD_LOGIC_1164.all;

entity redundant IS
port (input_a,input_b,input_s :in STD_LOGIC;
      out_y : out STD_LOGIC);
end redundant;

architecture archi of redundant is
--ip1_component--------
component ip1 is
port (a,b :in STD_LOGIC;
      y: out STD_LOGIC);
end component;

--ip2_component--------
component ip2 is
port (a,b :in STD_LOGIC;
      y: out STD_LOGIC);
end component;

--ip3_component--------
component ip3 is
port (a,b :in STD_LOGIC;
      y: out STD_LOGIC);
end component;

signal s1,s2,s3 : STD_LOGIC;
begin
 u1: ip1 port map (input_a,input_b,input_s,s1);
 u2: ip2 port map (input_a,input_b,input_s,s2);
 u3: ip3 port map (input_a,input_b,input_s,s3);
BEGIN
  PROCESS(s1,s2,s3)
  BEGIN
    if( (s1 = s2) and (s1 = s3) and (s2 = s3) then
      begin
        out_y<=s1; -- all ips working correctly
      end
    if( (s1 = s2) and (s1 \= s3) and (s2\= s3) then
      begin
        out_y<=s1; -- ip3 is not working correctly
      end
    if( (s1== s3) and (s1 \= s2) and (s2 \= s3) then
      begin
        out_y<=s1; -- ip2 is not working correctly
      end
    if( (s2== s3) and (s1 \= s2) and (s1 \= s3) then
      begin
        out_y<=s2; -- ip1 is not working correctly
      end
    END PROCESS;
end archi;


--IP1 from Vendor 1
ENTITY ip1 IS
  port(a,b,s:in std_logic;
      y :out std_logic);
END ENTITY ip1;
ARCHITECTURE ip1_a OF ip1 IS
BEGIN
    y <= ((not s) and a) or (s and b);
END ARCHITECTURE ip1_a;


--IP2 from Vendor 2
ENTITY ip2 IS
  port(a,b,s:in std_logic;
      y :out std_logic);
END ENTITY ip2;
ARCHITECTURE ip2_a OF ip2 IS
BEGIN
  PROCESS(a,b,s)
  BEGIN
    if( s= '0' ) then
      y <= a;
      else
      y <= b;
    end if;
  END PROCESS;
END ARCHITECTURE ip2_a;


--IP3 from Vendor 3
ENTITY ip3 IS
  port(a,b,s:in std_logic;
      y :out std_logic);
END ENTITY ip3;
ARCHITECTURE ip3_a OF ip3 IS
--components and signals BEFORE begin
component and2
    port(a,b:in std_logic; y:out std_logic);
end component;

component not1
    port(a:in std_logic; b: out std_logic);
end component;

component or2
    port(a,b:in std_logic; y:out std_logic);
end component;

signal a_and_nots, not_s, b_and_s : std_logic;

BEGIN
  Invert_S: not1 port map(s,not_s);
  An_And_Operation: and2 port map(a, not_s,  a_and_nots);
  Another_And_Operation: and2 port map(s,b,b_and_s);
  Final_Or_Operation: or2 port map(a_and_nots,b_and_s,y);
END ARCHITECTURE ip3_a;
```

Fig. 10.37 Verilog code to illustrate redundancy based solution

When accuracy of result is important, adhering to redundancy based solution is very important. More is the number of redundant IPs, better is the accuracy.

However, the main disadvantage of such redundancy based solutions is the huge overhead in area and power. Area and power overhead increases proportionally with the increase in the number of redundant IPs.

Hence, such solutions must be used only for critical sections of a design.

Such a technique is also useful for detecting malfunctioning IPs. When perfect results from two IPs are obtained, while a mismatch is found in the third one, then it can be inferred that the third IP is associated with some vulnerability and must be eradicated from the system.

10.7.2.2 Self Aware Agent Based Solution

Self aware agents are associated with monitoring of operations via side channel parameters like timing, power, etc. Based on deviations of these parameter from the normal range, a threat is detected.

Such self aware agents or modules are not part of the main program. These are normally designed separately by designers. However, they can schedule tasks on its host reconfigurable platform and ensure secure completion of the respective tasks.

Such self aware solutions can be applied either to centralized systems or distributed systems. For distributed systems, a self aware agent is associated with each host reconfiurable platform or a group of reconfigurable platform. The self aware agents can communicate among themselves and outsource task to one another, if necessary. For instance, if a self aware agent finds that its host reconfigurable platform has been compromised, then it can outsource the task to another agent, who will find some free time interval in the pre-existing schedule of its host reconfigurable platform to execute the task.

Mitigating confidentiality attacks via such self aware agents is well known for centralized systems. Mitigating availability attacks for both centralized and distributed systems can be performed via the self aware agents.

In this work, we will consider the side channel parameter, timing or delay to illustrate how confidentiality and availability attacks can be mitigated via self aware modules.

Mitigating Confidentiality Attacks

The self aware agents essentially monitor the time of operation, i.e. time taken from accepting the inputs to generating the outputs. If the operating time, increases or decreases via abruptly, then a threat may be associated. For instance, the threat demonstrated in Fig. 10.27.

To mitigate such threats, the self aware agent may use a random number generator to output the results at random time intervals. This will confuse the adversary and will not be able to decipher the secret.

Another possibility is to record the number of time units that are taken in generating the output. If more time units are taken in future, then the new value is recorded.

design.sv

```
module agent_conf(input clk, [3:0] ip_out,
                  output[3:0] out);

  reg [3:0] count=4'd0; // time units required for present operation
  reg [3:0] count_mem; // memory that stores the number of time units for max output

always @(posedge clk)
begin
  if (ip_out==4'd0)//output not generated, ip is working
    count<=count+4'd1;
  else //output generated, ip has completed working
    begin
      if (count==count_mem) //normal scenario
        out<=ip_out;
      if (count<count_mem) //may be threat
        begin
          #1;//wait for next clock cycle
          continue;
        end
      if (count>count_mem) //either threat or delay due to aging
      begin
        count_mem=count;
        out<=ip_out;
      end
    end
end
endmodule
```

Fig. 10.38 Verilog code to illustrate working of agent that can mitigate confidentiality attack

If at any instant, the number of time units taken are less than this recorded value, then the output is delayed till the recorded value is reached. In this mechanism, all the outputs are generated after fixed time units, regardless of the secret key value. This will prevent the adversary in deciphering the secret key.

For demonstration, we consider the example illustrated in Fig. 10.27. As evident from Fig. 10.29, based on key bit, output timing varied, i.e. when key bit was 0, output was generated without any timing delay, but when key bit was 1, a delay of around 2 time units in output was seen. Based on this, an adversary can decipher the secret key. Figure 10.38 depicts a verilog code for a self aware agent that mitigates such attack. This agent monitors the timing of the output. It has a counter which increments with each clock cycle, until the output is generated. When the output is generated, it checks the value of the count with the value of count_mem, i.e. the stored value of maximum time taken to generate the output. If these are same, then the scenario is normal, if count is less than count_mem, then there is a threat and via this anomaly, an adversary can decipher the secret key. Thus, delay is intentionally made by the agent till the value of count and and count_mem are equal and then the result is generated. Else if the value of count is more than the value of count_mem, then the additional delay may be a result of threat or due to aging. In that case, the new value of count is recorded in count_mem and the output is generated.

It is to be noted that the verilog code only demonstrates the working of the security module and this has to be incorporated on top of the verilog code represented in Fig. 10.28.

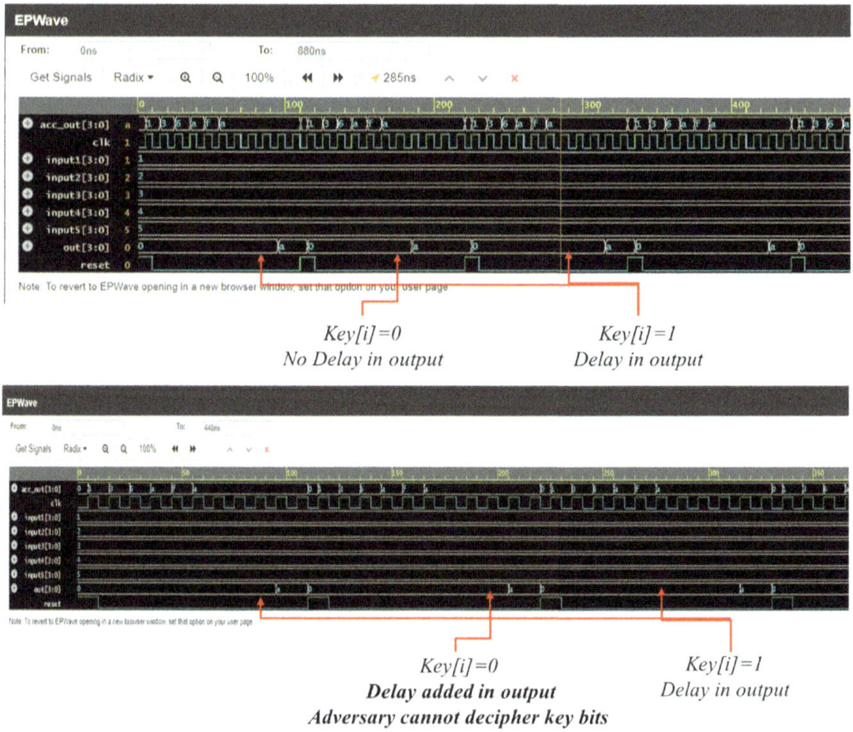

Fig. 10.39 Mitigation of confidentiality attack

Related results are shown in Fig. 10.39. The first figure shows variation of timing for key bits, while the second one depicts alignment of the timing for the outputs. This mitigates confidentiality attacks.

Mitigating Availability Attacks
For this portion, we consider a distributed computing environment, where multiple reconfigurable hardware platforms or FPGAs can be present. For demonstration, we consider a distributed system with three FPGAs. For each FPGA, a self aware security agent is associated. These agents can communicate among themselves via shared memories. The agents works based on the observe-decide-act (ODA) paradigm. This is depicted in Fig. 10.40.

Let us consider that a task executing in FPGA F1 needs 10 time units to complete, while FPGA F2 needs 5 time units to complete its task and 13 time units are taken by FPGA F3 to complete its associated task. Let the period of the system be of 15 time units, i.e. all the tasks repeats their operations after 15 time units. Let us also consider that the deadline for task in F1 is 10 time units, deadline for task in F2 is 5 time units and deadline for task in F3 is 13 time units. This is the normal scenario and depicted in Fig. 10.41.

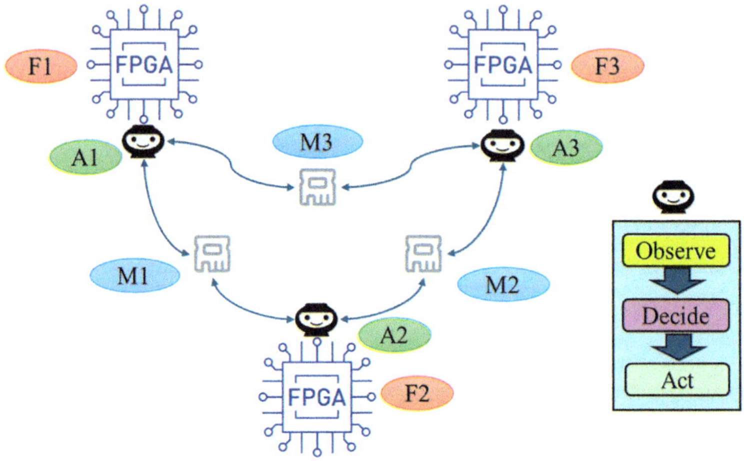

Fig. 10.40 Distributed scenario with multiple agents

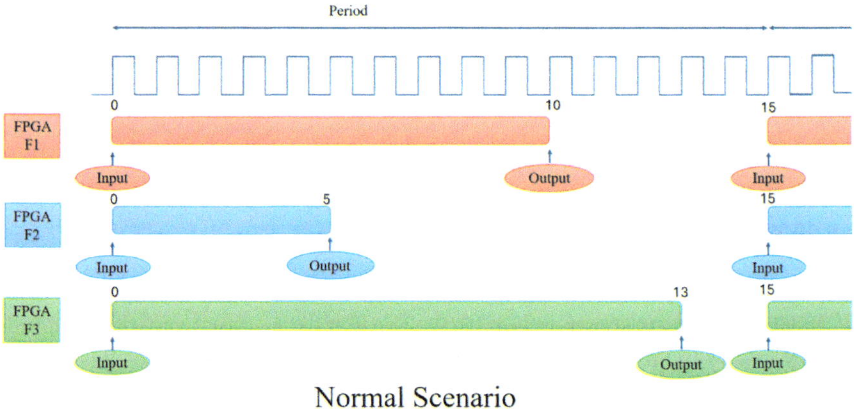

Fig. 10.41 Diagrammatic representation of task schedule during normal scenario

Let us consider that FPGA F1 is associated with an anomaly that prevents its completion before deadline. However, FPGA F2 and FPGA F3 operates perfectly. Such an anomalous scenario is depicted in Fig. 10.42.

On observing such an anomaly, agent A1 communicates with the other agents, A2 and A3 via shared memories, M1 and M3. This is associated with a memory write up of task details, i.e. arrival time, execution time, deadline and details of bitstream or reconfigurable intellectual property. A2 and A3 checks their host FPGA task schedules and if a suitable vacancy is found, accepts that task and completes it. For instance, A3 observes that there is a grace time of 2 time units before the next period and it has to do a five times speed up of operation. However, the deadline of task of F1 is 10 time units and cannot complete before the deadline. So, it does not accept the task.

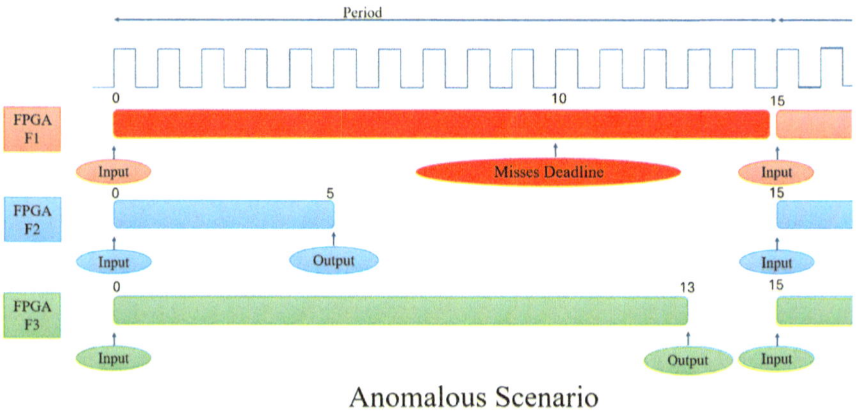

Fig. 10.42 Diagrammatic representation of task schedule during anomalous scenario

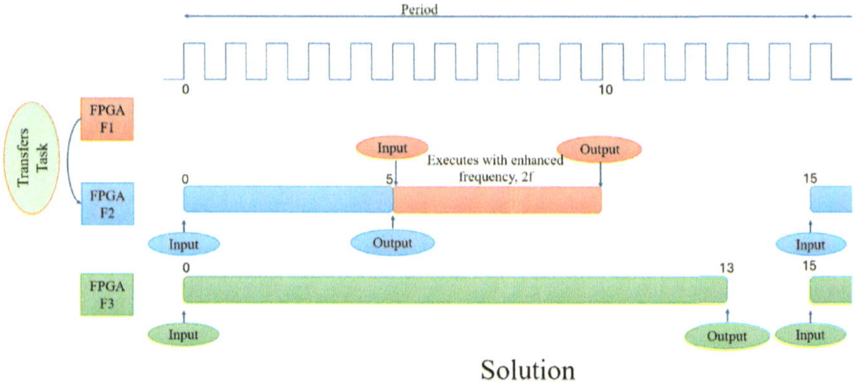

Fig. 10.43 Diagrammatic representation of solution

However, A2 finds that it can complete its host operation by 5 time units and has a grace time of 10 time units, before the next period begins. However, as the deadline of task of F1 is by 10 time units, hence, its effective grace time will be 5 time units and has to do a frequency speed up of 2 times. Hence, it accepts the task of F1 and executes that in its host FPGA F2 from 5 time units to 10 time units with two times the frequency of execution. Thus, task of F1 completes within its deadline. This is depicted in Fig. 10.43.

Verilog code for working of a self aware agent is depicted in Fig. 10.44. It essentially comprises two parts. The first part is associated with checking operations of its host FPGA and the second part involves checking the shared memories and if any task is sent, then deploying them in its host FPGA.

For the first part, a counter is deployed from the time its input arrives to the time the output is generated. However, if the value of the counter exceeds the value stored

design.sv

SV/Verilog Design

```
module security_agent2(input clk, reset, [3:0] fin, [3:0] fout,
                        output [3:0] m1, [3:0] m2);
  //fin and fout depicts the task input and task output for host FPGA
  //m1 and m2 depicts the shared memories for communicating with other two agents of the system
  reg [3:0] counter; //counter counts the number of clock cycles or time units required for present task execution
  reg [3:0] count_mem=4'd5; //count_mem stores the number of clock cycles or time units required for task execution
  reg [3:0] period= 4'd15;
  //for simplicity, it is assumed that task input and task output cannot be zero and a non-zero value for task input
(fin) denotes arrival of task and a non-zero value for task output (fout) denotes completion of execution

always @(posedge clk or posedge reset)
begin

  //checking host operation
  if(reset) //all shared memories are reset, along with task input and task output, i.e. fin and fout respectively
    begin
      m1=4'd0;m=4'd0;
      fin=4'd0; fout=4'd0;
    end

  if(fin!==4'b0 and fout==4'd0) //if task input comes, but task output does not come, i.e. execution in place
    begin
      counter<=counter+4'd1;
      if(counter>count_mem) //if present time exceeds the recorded time, then other agents have to be communicated
        begin
          //setting of m1 and m2 with its host inputs, i.e. fin, depicts start of communication with other agents
          m1<=fin;
          m2<=fin;
          break;
        end
    end

    //analyzing host for executing operations of others
  if (m1!==0) //F1 task has not completed within time
    begin
      #5;   //complete its own task
      DCM<=2; //enhance frequency by 2 times, via signaling dynamic clock management (DCM) to enhance frequency twice
      fin<=m1; //deploy task input of F1
    end
  if (m2!==0) //F3 task has not completed within time
    begin
      #5;//complete its own task
      DCM<=2;//enhance frequency by 2 times, via signaling dynamic clock management (DCM) to enhance frequency twice
      fin<=m2;//deploy task input of F3
    end

end
endmodule
```

Fig. 10.44 Verilog code for demonstrating working of a self aware agent in a distributed environment

in memory, i.e. the normal time it takes to complete its operation, then it infers that an anomaly can be associated. In such a case, it writes its input to the shared memories.

For the second part, it checks which shared memory is set. In such a scenario, after its operation completes, i.e. after a delay of five time units, it enhances its dynamic clock management (DCM) module by the related amount, i.e. two times. This DCM is a separate portion in the FPGA that enhances its frequency of operation. Alongside, it deploys the input that is present in the related shared memory.

Figure 10.44 demonstrates working of agent A2, that is associated with F2, which executes a task for 5 time units. M1 is set if F1 is associated with a delay and M2 is set if F3 is associated with a delay.

Waveform for the normal scenario, anomalous scenario and solution is depicted in Fig. 10.45, Fig. 10.46 and Fig. 10.47 respectively. For simplicity, we consider that the task output is same as task input, but F1 takes 10 time units, F2 takes 5 time units and F3 takes 13 time units for execution. Figure 10.45 depicts the output of F1, F2 and F3 at appropriate times. In Fig. 10.46, there is no output of F1 and hence, availability of the system is affected. Finally, Fig. 10.47 depicts the solution, where output of F1 is produced in F2, after completion of task of F2.

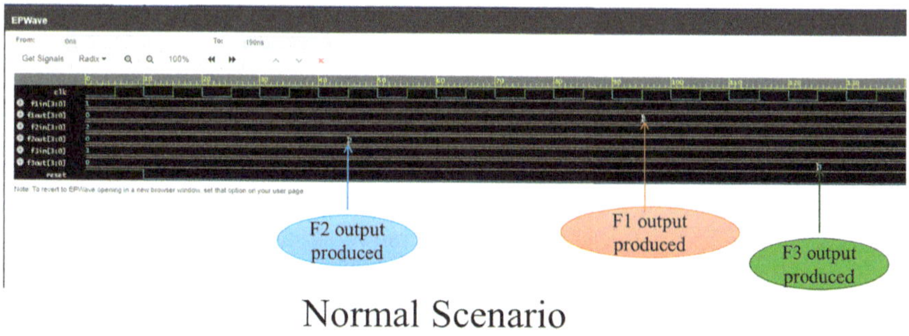

Fig. 10.45 Waveform depicting task schedule during normal scenario

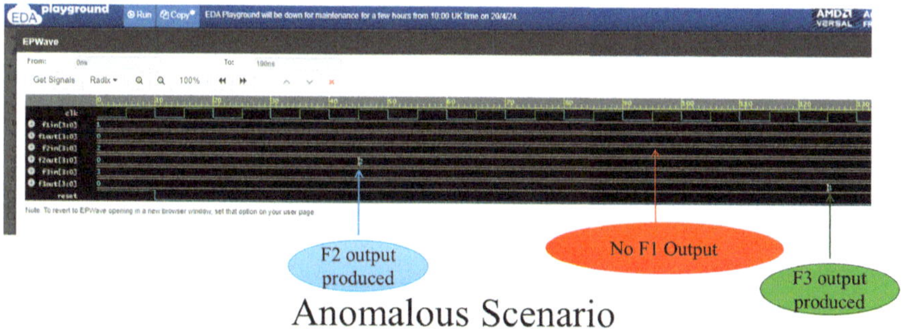

Fig. 10.46 Waveform depicting task schedule during anomalous scenario

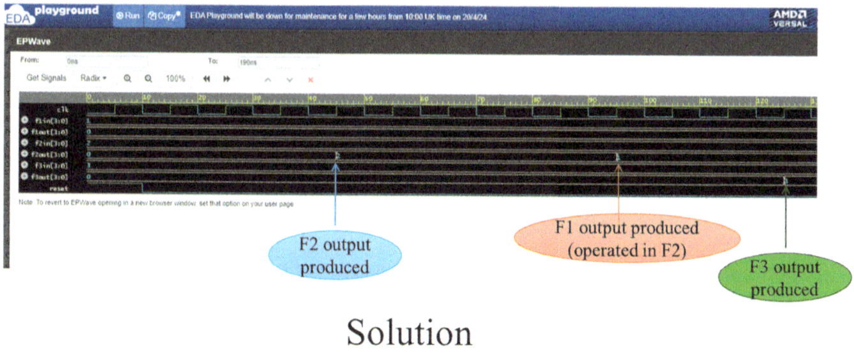

Fig. 10.47 Waveform depicting task schedule during solution

Questions

1. Consider the following switch light logic problem. There are two switches and a bulb. The bulb will glow only if both the switches are on. Explain what type of logic operation will be associated to solve the following problem and write a HDL program for it, and generate waveform to demonstrate its working.
2. State the different types of HDL programming and demonstrate them for developing a 3 to 8 line decoder.
3. Develop a FSM Accumulator with HDL (VHDL or Verilog) and explain its working with the help of a waveform.
4. Depict by writing a HDL program how a HTH may affect the working of the FSM.
5. Demonstrate with an HDL program how XOR based encoding and decoding can be performed.
6. Depict how confidentiality attack can be carried out by an HTH on the XOR based encoding for the previous question.
7. Explain with HDL programming, how finite and infinite delays can be made on any functional unit.
8. Write a HDL program to demonstrate how key based obfuscation can be carried out.
9. Write a HDL program to demonstrate how triple modular redundancy can be useful for a critical functional unit.
10. Write HDL programs to demonstrate working of self aware agents for (i) mitigating confidentiality attacks (ii) mitigating availability attacks.

References

Bhasker J (1994) A VHDL primer (revised ed). Prentice-Hall, Inc

Palnitkar S (2003) Verilog HDL: a guide to digital design and synthesis, vol 1. Prentice Hall Professional

SPRINGER NATURE

GPSR Compliance

The European Union's (EU) General Product Safety Regulation (GPSR) is a set of rules that requires consumer products to be safe and our obligations to ensure this.

If you have any concerns about our products, you can contact us on ProductSafety@springernature.com

In case Publisher is established outside the EU, the EU authorized representative is:

Springer Nature Customer Service Center GmbH
Europaplatz 3
69115 Heidelberg, Germany

The manufacturer's authorised representative in the EU is Springer Nature Customer Service Centre GmbH, Europaplatz 3, 69115 Heidelberg, Germany. If you have any concerns regarding our products, please contact ProductSafety@springernature.com

Printed and bound by CPI Group (UK) Ltd, Croydon, CR0 4YY

30/09/2026

02225563-0001